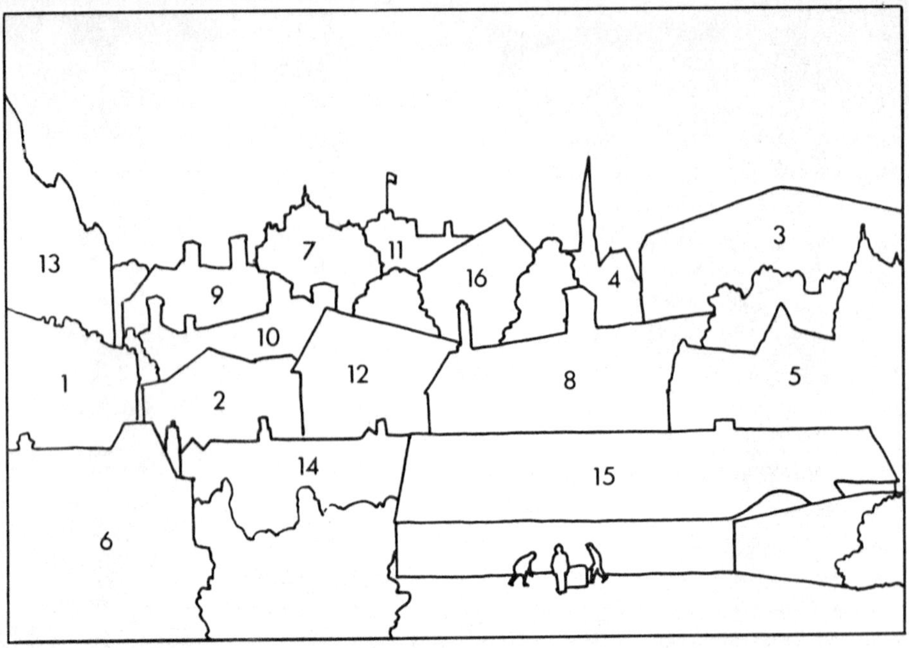

Buildings illustrated on cover.

1. Doncaster Mansion House. 1748.
2. The Dispensary, Doncaster. 1793.
3. The Market Hall, Doncaster. 1847.
4. The Cemetery Chapels, Doncaster. 1855.
5. The Infirmary, Doncaster. 1865.
6. The Library and School of Art, Doncaster. 1889.
7. York City and County Bank, Doncaster. 1898.
8. Principal's House, Kings Manor, York. 1900.
9. Thorp Underwood Hall. 1902.
10. 'Bishopsbarns', St. George's Place, York. 1905.
11. Keldy Castle, Pickering. 1910.
12. St. Luke's Church, York. 1910.
13. T. Cooke & Sons Factory, Bishophill, York. 1916.
14. 'Goddards', Tadcaster Road, York. 1926.
15. Church/Hall of James the Deacon, York. 1955.
16. Big Hall, St. Peter's School, York. 1960.

The cover design is a lithographic reproduction from the original painting by Russell P A.R.C.A. It was specially commissioned for this occasion by William Anelay Ltd. who wis express their thanks to Mr. Platt.

A SKILFUL MASTER-BUILDER

The continuing story of a Yorkshire family business
Craftsmen for seven generations

by Revd. H. E. C. Stapleton, M.A., F.S.A.
Editor

George G. Pace, C.V.O., M.A., F.S.A., F.R.I.B.A.

and James E. Day

Published by
William Anelay Ltd., Murton Way, Osbaldwick, York
in association with
William Sessions Ltd., The Ebor Press, York

This book has been issued as a contribution towards the objectives of European Architectural Heritage Year, 1975

ISBN 0 900657 34 0

Printed in England in 1975
in Times Roman typeface
by H. Morley & Sons (York) Ltd., York, England

FOREWORD

by the Marquess of Normanby

I AM very pleased and honoured to have been asked to write a foreword to this excellent history of the Anelay family and firm, William Anelay Ltd.

All who read this most interesting account will be filled with admiration at the very able way in which Mr. James E. Day, Mr. George G. Pace and the Revd. H. E. C. Stapleton have approached their subject from different angles, and so have provided a very complete history of the family and of the firm set in their contemporary background through nearly 250 years.

A high standard of craftsmanship and work would appear to have been set by the members of the Anelay family from the very first – this tradition has been maintained in a most astonishing and striking manner.

I suppose that it is generally true to say that quality and efficiency are best judged by those who have sampled them personally. As a client of Messrs. Anelay I know that their standard of work and workmanship is outstanding. One is led to question how this has been maintained and improved, and how new methods have been smoothly introduced and problems successfully solved in each generation. As an onlooker, I have often thought that this must be mainly due to the very friendly, personal relations which exist within the firm, and between the firm and its clients. This is fostered just as much by those who work for William Anelay Ltd. as by those who own or direct it.

I join with very many others who wish this firm well, in hoping that this account will be enjoyed by a wide public – it will, I am quite sure, inspire and interest those who read it.

NORMANBY

MULGRAVE CASTLE
WHITBY

CONTENTS

ILLUSTRATIONS

TABLES

ACKNOWLEDGEMENTS

IN 1971, Mr. Peter Anelay asked me to write the history of the Anelay family and the firm, William Anelay Ltd.; at the same time he invited Mr. Pace to write a paper on the place of the firm in the history of Building. I had little to go on except Hatfield's *Doncaster*, the registers of Doncaster Parish Church and a few family archives. Within a few months, however, I managed to present him with a brief typescript which I felt contained all there was to know. But soon the fruits of much research by the Doncaster Central Library began to reach me and in due course Mr. Day revealed himself as a mine of information about Doncaster, the Anelay family and much else besides. As a result, instead of the originally planned two parts, there are now four, as will be seen on the Contents page.

My gratitude therefore must first go to Mr. Anelay for giving me the privilege of compiling this book, for his constant assistance and keen encouragement. I must also thank the Marquess of Normanby for his introduction; Mr. Pace and Mr. Day for their papers; the many clergy who have patiently assisted our investigation of the parish registers, especially the former Vicar of Doncaster the Revd. M. Jackson, the Vicar of Arksey the Revd. W. A. Stratford, and the Vicar of Snaith the Revd. C. J. Hodge; the Reference Librarians of Doncaster Central Library (Mrs. M. Jackson and Mr. C. Howarth), and Mr. Maurice Smith of the York City Library; Mr. Edmund Cooney of the University of York; the Superintendent Registrar of Doncaster, Mr. F. Holmes; the Editor of the *Doncaster Gazette*, Mr. K. Dutton; Mr. Garry Swann, Senior Medical Photographer, Doncaster Royal Infirmary; Mr. V. Douglas Knox, D.F.C., Mr. W. H. Gordon Smith, Dr. George Redmonds and Miss A. L. Dawson.

HENRY STAPLETON,

January, 1975.

THE SPONSORS of this volume, William Anelay Ltd., wish to express their sincere appreciation, not only to those mentioned above, but also to the Revd. Henry Stapleton, Rector of Skelton, who, as Editor of this history has skilfully synthesised the four parts of this book, so as to present a vivid picture of many facets of industrial and social life in Doncaster, York and Yorkshire from Georgian times to the present day.

The illustrations on pages 5 and 30 are reproducd from *1800 Woodcuts by Thomas Bewick and his school*, edited by Blanche Cirker, published by Dover Publications, Inc., New York, 1962.

INTRODUCTION

'YOU are one of the few people who seem to care for beauty and workmanship. I know Mr. Comper has the same feelings towards you as I have.' So wrote the Revd. W. M. Tatham to Thomas Anelay VI after the completion of the repair of Cantley Parish Church. These words implicitly commend that subtle trinity of relationship between client, architect and contractor, a quality that is distinctive of William Anelay Ltd. Though written in 1934 this description would equally well describe the founder in 1740, John Thompson 'a clever and skilful master builder'* or any of the seven generations of Anelays, who were to succeed to his business.

The survival of any business for a century is notable enough. That one concerned with building should still be in existence after two and a quarter centuries is remarkable. That the same family still runs the concern is exceptional, almost unique. From father to son the management of the firm has been passed down in unbroken succession. Continuous too has been a tradition of workmanship and integrity. First the term 'a Thompson job' and later 'an Anelay job' has become proverbial for a job carefully, conscientiously and well done.

* C. W. Hatfield, *Historical Notices of Doncaster*, Vol. II, page 290.

THE ANELAY GENEALOGY

Thomas Anelay of Heck, Gowdall, Snaith
d. 1695

Thomas Anelay I
b. 1675 Snaith
d. 1736 Snaith

John Thompson m. 1740 Catherine Barwick
b. 1714 — b. 1718
d. 1771 Doncaster — d. 1798 Doncaster

John Anelay of Pollington m. 1734 Hannah Johnson
b. 1710 Snaith
d. 1765 Snaith

Ann Thompson m. 1760 John Anelay
d. 1812 Snaith — b. 1736 Snaith
d. 1761 Snaith

(John Anelay's wife, Ann Thompson,
was either daughter or neice of
John Thompson)

Thomas Anelay II m. Sarah of Jump, nr. Barnsley
b. 1739 Pollington — b. 1733
d. 1798 Doncaster — d. 1812 Doncaster

Thomas Anelay III m. 1799 Hannah Doncaster
b. 1764 Southwark — b. 1767
d. 1848 Doncaster — d. 1841 Doncaster

Thomas Anelay IV m. Sarah Hatfield
b. 1805 Doncaster — b. 1801 Misson
d. 1886 Grimsby — d. 1880 Doncaster

Charles Augustus Anelay
b. 1808 Doncaster
d. 1876 Stainforth

Thomas Anelay V
b. 1833 Doncaster
d. 1917 Doncaster

William Anelay I m. 1865 Clara Milan
b. 1841 Doncaster — b. 1845 Doncaster
d. 1918 Doncaster — d. 1925 Doncaster

Thomas Anelay VI m. 1892 Jesse Frances Whitehead
b. 1866 Doncaster — b. 1868 Friern Barnett
d. 1942 Harrogate — d. 1945 Scarborough

Thomas Anelay VII
b. 1896 Doncaster
d. 1966 Doncaster

William Anelay II m. 1925 Norah Daisy Wilson
b. 1900 York — b. 1900 Brighton

William Peter Anelay m. 1956 Jean Baines
b. 1927 York — b. 1932 York

Charles Thomas Anelay
b. 1959 York

Helen Felicity Anelay
b. 1961 York

Elizabeth Catherine Anelay
b. 1963 York

Thomas Anelay IV
(1805-1886)

Thomas Anelay V
(1833-1917)

William Anelay I
(1841-1918)

Thomas Anelay VI
(1866-1942)

William Anelay II
(1900-)

William Peter Anelay
(1927-)

2 *The Mansion House, Doncaster, built between 1745 and 1748 with brickwork by John Thompson. Architect, James Paine. In 1801 Paine's pediment was removed and a new attic storey was built by Thomas III. Architect, William Lindley. They did further work in 1806.*

3 (above) *Doncaster in 1802, looking south down French Gate and High Street on the Great North Road. In the foreground is Marsh Gate which John Thompson, with Henry Walker, raised on arches in 1753 to prevent flooding. The commemorative plaque is in Cusworth Hall Museum.*

4 (below) *The Doncaster Dispensary, French Gate (now demolished) built by Thomas III in association with George Hewitt in 1793 at a cost of £665.0s. 8d. Architect, William Lindley.*

TO BRICKLAYERS.

FIVE JOURNEYMEN BRICKLAYERS may meet with Employment by immediate Application to Mr. THOMAS ANELAY, of Doncaſter.
☞ Wages according to Merit.

5 (above) *Rossington Rectory, Doncaster, built 'as a school house and school room' in 1801 by Thomas III and converted by him into a Rectory in 1804.*

6 (centre left) *An advertisement in the* Doncaster, Nottingham and Lincoln Gazette, *17 May, 1805, inserted by Thomas III for five bricklayers.*

7 (below left) *St. Leger Day at Doncaster Racecourse showing John Carr's Grandstand which Thomas III altered in 1804, and the elegant Judges' and Stewards' Stand which he built in 1805.*

PART ONE: A HISTORY OF THE ANELAY FAMILY AND FIRM

CHAPTER I

JOHN THOMPSON (1714-1771)

TRADITION has it that John Thompson, of Doncaster, founded his firm in 1747. Recent evidence has, in fact, proved that he was doing work for Doncaster Corporation as early as October, 1740. The following receipt is contained in a parcel of miscellaneous bills amongst the Corporation Archives.[1]

March 8 1740	
Brickwork for Mr. Heaton	
getin a load of Lime and sand into Shroo [Shoe] Markit and Slekin it	6
3 days & a half a bucher Shamels [Shambles]	5 10
a laber 3 days	3 0
days at ye awnes houses by y[e] [Alms-houses] Corch yard & in y[e] moudlens [Churchyard & Magdalens]	6 8
a laber 7 dys	7 0
5 days at y[e] houses by ye hal cross [Hall Cross]	8 4
a laber 5 days	5 0
3 days pavein in fishergate	5 0
a laber 3 days	3 0
	2 4 4

John Thompson Bill
March y[e] 10 1741
Reced y[e] contents by me

John Thompson

According to the inscription on his gravestone in the churchyard of the Parish Church, John Thompson was born in 1714. There is no certainty about the identity of his father although there is a baptismal entry for that year of one, John, who was the son of John Thompson, a clockmaker. He married Catherine Barwick (1718-1798)[2] but apparently no male children survived infancy. John Thompson had therefore to look beyond his immediate family for a partner in his business.

[1] *See Notes commencing on page 22*

In 1740 he had been a witness at the Snaith marriage of his daughter or niece, Ann Thompson, to John Anelay II. Sadly, the bridegroom died childless within the year. However, his brother Thomas Anelay II was already in the building trade and in 1767 John Thompson invited him to join his firm and devised it to him at his death.[3]

The history of Doncaster goes back to Roman times. Its strategic position on the Great North Road caused its development into a town of some importance, prior to the coming of industry and the railways. Its wealth was reflected in the size of municipal estates and the dignity of the Mansion House; York was the only other place to possess a building so-called outside London. In the 18th century wealthy families built houses in the park-like country around, coming to the town for social occasions, of which the race week in September was the height. The St. Leger was first run in 1776 and has been run every year since with the exception of the war years, when the venue was removed to Newmarket. Over the town soared the tower of the Parish Church, grievously devastated by a disastrous fire in 1853 and later rebuilt by Sir George Gilbert Scott.

Work on the construction of the Doncaster Mansion House began in 1744 but work was delayed by quarrels within the Corporation and by the Rising of the Young Pretender in 1745. An army of 6,000 were encamped on the Wheatley Hills and John Wesley records in his Journal:[4]

> Thursday October 10, 1745. We lay at Doncaster, nothing pleased with the drunken, cursing, swearing soldiers, who surrounded us on every side.

John Thompson was apparently a bricklayer at the Mansion House in 1747. Hatfield says his reputation 'was well sustained in the erection of the Mansion House and an inspection of the extensive cellars leaves no doubt on this point'.[5] There were disputes respecting former contracts and work done in respect of the building and a writ was served on the original committee.[6] Perhaps that was the reason for the limit of Thompson's work. The architect was James Paine, of Wragby, who subsequently did work at Cusworth Hall, Serlby Hall near Bawtry and in 1752 at Cowick for Viscount Downe.

On 13th October, 1748, the following resolution of the Corporation was recorded:[7]

> Ordered that John Thompson do pave ye Corporation's pavement and that he be allowed £18 per annum for the first three years and £15 for ye next four for keeping ye same in good repair and that he enter to it from Mich'mas day last.

In 1753, the Corporation, in order to make the passage of traffic on the Great North Road safer, raised a loan of £300 to pay for a causeway between Friar's Bridge and Navigation Bridge. Thompson, with his high reputation as 'a clever and skilful master builder, proverbial for his good work in the country around', was given the contract of raising the road Marshgate and of protecting it by a substantial stone wall.[8]

In view of Thomas Anelay II's reputed interest in field sports, it is interesting to read the Corporation Minute of 4th April, 1768:

> Ordered that the Corporation Steward do give notice to John Thompson to quit possession of the Barn in Laithgates belonging to the Corporation on or before May day next, and that the same be put in repair by the Steward, and used for a Dog-kennell for their hounds.[9]

John Thompson died in 1771 and was laid to rest in the grave which would later receive his wife and their five children. Hatfield thus describes his resting place:

> He slumbers beneath the shade of the magnificent south chapel of St. George's Church—an appropriate resting place for one so industrious and honest in his calling.[10]

CHAPTER II

ANELAY ORIGINS

ALTHOUGH there are many different spellings of the family surname, for example, Anley or Ainley or Hanley, Dr. George Redmonds[1] believes that the family comes from the same place and district called 'Ainleys', outside Elland on the Huddersfield Road in the West Riding of Yorkshire. Of the origin of the name he writes:

> The final part of the place-name offers no problem and is clearly from the Old English word 'leah' that is 'a clearing in a wood'—and there is still wood land clinging to the steep hillside known as Ainleys. The latest theory regarding the first element is that it was originally an Old Norse personal name best interpreted as 'Agwind.' The meaning of the place-name would, therefore, be Agwind's clearing.[2]

In the Middle Ages Elland was a chapelry within the large parish of Halifax, and the family name is first found in a document concerned with the small township of Fixby. Thomas de Auenlay and William de Avenlay sign as witnesses to an agreement of 1292[3] involving Henry de Fekesby and Adam de Rastric. It is remarkable that these two Christian names, Thomas and William, recur time and again through the generations right up to the present day. Eighty years later in the Poll Tax Register of Elland there is a record of Johne de Hanley and his wife and his children John and Isabella.[4] During the 15th century, an Anelay was invariably a member of the jury of the local manor court. One Richard Anelay followed another and it is hard to discern the relationship with any accuracy. The earliest extant will,[5] that of Gilbert Hanley of Elland, of 1539, however, gives a clearer picture of one family:

> November 6th. To be buried in the churche yerde of Oure Ladie in Elande. Itm. [Item]. I give to the forsaid churche 3 shillings and 4 pence which is in the handes of John my broder. To Robert Hanley, my sone a blewe jacket. To Agnes Smythe my daughter, a violet jacket without sleves.
>
> To Jennet my wif the termes of my farmholde whiche I holde by indentures, so longe as she is unmaried; and if it fortone [fortune, i.e. happen] that the forsaid Jennet do marie, I will that Richarde my sone have the residue of the forsaid termes. I will that Jennet my wif give to Thomas, Michaell and Margaret my childer [children], euere one of them 6 shillings and 8 pence. Also I will that Richarde my sone give to euere one of the forsaid childer 10 shillings over and beside the childer parte and this soms [these sums] to be done at all such tymes as the frendes of the forsaid childer my wif and Richarde my sone whome I make myne executors. These witnes, John Brokbanke, prest, Ric. Hanely, Jeffray Romsden and John Smythe.

In the second early reference, it would seem that Richard is again the eldest son. He may be the 'Richard Anley of Ouldlinley' who had a stall in the chancel of Elland Church adjacent to that of 'John Hanson for overwoodhouse'.[6] Another Thomas of Stainford, in 1589, was one of the many Anelays who served as churchwardens at Elland in the 16th and 17th centuries.[7]

It is in the 16th century that the Anelay family first appears in York. In 1548/9, Richard Aynlay[8] or Ayneley,[9] vintner, was made a Freeman of the City. Apparently he lived in the parish of St. Martin, Coney Street.[10] He had seven children. The first-born, Richard, did not survive infancy and was buried in 1565; Elizabeth, born in 1559, died 1567; John was born and died the same year 1563. Alice was born in 1560, Mary 1564, Thomas 1565 and Henry 1566. Henry later lived in Gillygate, York,[11] and, by his wife Francis, had a daughter Mary and an apparently brief lived son John, possibly named after his equally hapless uncle. The entries in the St. Olave's registers are interesting, not only for the delightful spelling of the street where they lived and the 'ay' in Francis surname but also because three centuries later Thomas Anelay VI and then William II were to come and live within the same parish.

August 1598 Marie Ayneley filia Henrici Ayneley de Jellygait babtizata xxvii

1602 Johannes filius Henrici Aynley de Jellygait mort et sep primo die Julii

1610 gilligait, francis aynlay wyef of Henry Buryed iiii march

Henry was a haberdasher and became a freeman of York in 1590, his father being described as 'gentleman'.[12]

Richard Anelay, the vintner of Coney Street, could be identified with the owner of the stall in Elland church especially as there are records of litigation between Hansons and Thomas Anelay in the next generation, and Thomas was the name of Richard's eldest surviving son,[13] but there is no conclusive evidence and Richard is probably the 'notary' buried at St. Martin's in 1575; and no descendants have yet been traced.

York was but one of the places to which the family from Elland went. Others can be found at Wath on Dearne,[14] Hooton Pagnell[15] and Brodsworth.[16] But the branch from which the Doncaster family springs came from the ancient parish of Snaith[17] between Doncaster and Selby. Until smaller units were formed in the last century it comprised an area of nearly 35,000 acres[18] including an area of townships such as Cowick, Gowdall and Pollington where various relatives lived. A considerable variety of spelling of the name persists; but the consistency in the use of Christian names makes easier the task of sorting out the various families and also gives a strong impression that the Doncaster Anelays look back to their ancestry in Elland. The first Anelay by the name of John appears in the Snaith Registers in 1602. A Thomas is first mentioned in 1675, Charles in 1696, William in 1698, Richard in 1709 and Elizabeth in 1710.

The earliest of all the Anelay entries at Snaith is Nicholas Anley who died in 1602, six days after his infant son, John. It is tempting to see Nicholas as the ancestor of the family and Thomas who was buried in 1695 as his grandson but there is no direct link. This Thomas Anelay of Heck, however, is the ancestor of this branch, the father of Thomas I of Pollington and Charles of Heck. Charles married, on 15th April, 1696, a lady delighting in the name of Saltmarsh Toppom. Such an unusual Christian name may have been derived from her mother's origin at Saltmarsh near Howden in the East Riding. The name was repeated in subsequent generations; their second daughter was so called at her baptism at Snaith on 9th March, 1703. In 1762 a great granddaughter, Saltmarsh Ainsley, married Thomas Darbison of Newland in the parish of Drax at Carlton church.[19] The baptisms of their eighth son Robert and tenth son, Ebinazar, are also recorded in the same register.

Thomas I's children were John I, baptised at Snaith 4th February, 1710, Elizabeth, 14th April, 1713, Charles, 9th July, 1717, and William, 28th September, 1721. Charles was accidentally drowned in 1799, being described in the register as a labourer. An unnamed child who was buried on 29th July, 1712, may have been the infant Thomas, baptised on 17th May, 1709. Hatfield's *Historical Notices of Doncaster*, 1861, records that John I was a tenant of Lord Downe on his estate at Pollington.[20] Hatfield is suspect for his historical flights of fancy but wherever in this account it has been possible to check, no fault has been found, although his interpretation of the facts is sometimes questionable. There is a possibility despite the difference in age that he was a friend of Thomas V, the Borough Remembrancer, whose mother was Sarah Hatfield. According to the report in the local press the historian was laid to rest in an Anelay coffin.[21]

John I's family were also all christened at Snaith: John II 2nd September, 1736, Thomas II 27th August, 1739, Charles 16th May, 1742, Mary 10th February, 1744, Betty 21st February, 1747, and William 26th August, 1749. In 1754, a Mr. Anelay is recorded in the Doncaster Corporation Rate Book as being resident on Goose Hill 'over Richard Pickman's shop';[22] but this is unlikely to have been John Anelay I from Pollington, as his burial is recorded at Snaith in 1759 or 1764.

Three of John I's sons were to become blacksmiths: John II at Snaith, where he was married to John Thompson's daughter or neice, Ann; Charles at Mill Bridge, Doncaster, and William at Pollington. The second daughter, Betty, married Michael Oxley, a Doncaster builder,[22] who appears on the Anelay payroll in October, 1773.[24] The second son Thomas II (1739-1798) went up to London and lived in Tooley's Gate, Southwark.[25] This street is shown as 'Tooly's Gate' on Rocque's Map of London, 1746. It is now represented by the dog-leg called English Grounds which lies near London Bridge Station on the north side of Tooley Street between Morgan's Lane and Battle Bridge Lane. To the parish church of St. Olave, the 'bricklayer' Thomas II, and his wife Sarah brought Thomas III for christening on 29th July, 1764, and Elizabeth on 4th January, 1767. The reason for his sojourn in Southwark is not known. He may perhaps have hoped to follow the success of his cousin Samuel,[26] the son of Richard Anley of Stainford in the parish of Hatfield, who had become a tailor in the City. However, in 1767 he returned to Yorkshire in order to work with his master-builder 'uncle' John Thompson in Doncaster.[27]

Builder

CHAPTER III

THOMAS ANELAY II (1739-1798)

JOHN THOMPSON died in 1771. The survival of Thomas Anelay II's Day Book 1772-1774, and descriptions in Hatfield, which may be family reminiscences, enable the painting of a portrait of the successor of the 'skilful master builder'. The very first entry is significant:

> April the 28 1772 for Corporation at Hextrop [Hexthorpe] & [April] 29: 2 days
> ½ Shackleton & Scales[1]

Thomas II and his two men were already engaged on work for the Corporation. There are numerous other similar entries varying from 'leading' bricks to the building of causeways; sinking a well, bridge repairs and demolition of almshouses. So much so that one is tempted to see the firm almost as the Corporation's Works Department. In 1773 the firm undertook the removal of the Wheat Cross,[2] a single pillar surmounted by a stone ball and a circle of flags at the base of which the wheat sacks were pitched, and replaced it with an ornamental wall. More substantial enterprises undertaken were the erection in 1779 of the second Doncaster Jail in St. Sepulchre Gate under the architect, William Lindley[3] and in 1794 the Dispensary,[4] completed at a cost of £665. There were alterations, too, to houses large and small.[5]

21 [August 1772]	at Mr. Holms Stucoing Dinen Room Plaster and sand 1s 6d and 1 Di Slack & Son
29 [March 1773]	at Mr. Kellam Sommer House ¼ Day Self 1 Day Chappell Tatershell & over Brick 1400 Common & 400 Drest Bricks
30	1 Day Chapl Tatershell Scaffild Boards
31	Paid 5s for Timber for Roof Paid Palmer & 1 Day Chapl & Tatershell for leading 13s 6d
April 3	1 Day Chapl Tatershell

The name of Kellam or Kilham will occur again in a different context. Perhaps what was done at Sir George Cook's, a mansion at Wheatley,[6] near Doncaster, were rehearsals for the works on large houses they were later to undertake.

The 18th century was not an age when the established church displayed much vigour in the erection of churches. In fact, the first ecclesiastical building mentioned is the Dissenting Chapel on the west side of Hallgate of which Mr. Scott was Minister.

> 9 [June 1772] for Mr. Scott at Meetin hous.

Thomas did much work at Rossington.[7] The following entries from the Day Book tell the sequence of events:

18 [April 1774] at Rectory at Rossington

10 1 Day Shaklton 5 day Lad & Overton

9 [May 1774] Repairing Vestry parte of Rossington Church

14 At Church Mending Steple & Leds collering [collaring] do.

1 Do Lad firth

23 [May 1774] at Rossington Church 1 day Lad & firth[7]

In 1778 he was to build the Court House in the village.[8]

The Day Book records not only what work was carried out each day and the materials but also the wages they were paid. For an analysis of these records see pages 59 and 60. Thomas II seems to have paid himself the same as his 'lad' and this amounted to less than each of the 13 employees earned. Hatfield says[9] 'he was a stickler for punctuality. For the election of a mayor, his workmen kept it as a half holiday and they made merry'. With scrupulous care Thomas II records the borrowing and repayment of sums from both his 'Ant Thompson', and his grocer neighbour.[10]

He augmented his income by the letting of lodgings[11] and also by an occasional valuation of property. In 1784 he took on the responsibility of being gamekeeper on the Corporation's estates at Wheatley cum Sandall. Some doubt has been cast as to whether it would be possible for a builder to desert his business for country pursuits. That he was able to do this successfully may be deduced from the fact that in 1779 he was lending the Corporation (his employers) £100.[12] The status of the post of gamekeeper is evidenced by the annual salary of 50 guineas later to be voted by the Corporation to the holder of the office in 1807.[13] Thomas II received but two guineas.[14] His social position however, is perhaps better indicated in the Corporation accounts where he is punctiliously referred to as 'Mr. Thos Anelay'.

If the transmission of an interest in antiquities from one generation to another is adequate evidence there could be some probability that Thomas II was acquainted with the eminent Dr. Miller (1737-1807), the organist of the parish church and author of the *History of Doncaster*. Hatfield wrote that they went on shooting expeditions together but this is apocryphal.[15] Some authorities even doubt if the musician ever possessed a gun! The firm carried out work on Dr. Miller's property 1798-1804. The second account book has certain pages at the back labelled Sand Book, in which there is a record of the leading of 35 loads of sand from the doctor's quarry between March 1798 and December 1799.

Earlier mention has been made of the place of the Races in the life of the town. Thomas II enjoyed the races too. In February 1777 it was ordered that 'one hundred and fifty Tickets for the Admission of persons into the Stand to be Erected upon the intended race course to be offered for Sale, at Seven Guineas each Ticket, and that the Noblemen and Gentlemen that live in the Town and Neighbourhood have the first refusal thereof'.[16] Thomas II was allotted Silver Ticket No. 59, inscribed T. Ainley, which remained in the family possession until 1948. He was at the Doncaster Races in 1787; his marked race-card is still preserved. The firm worked on alterations to the Grandstand in 1804, the construction of the Judges and Stewards Stands in 1805, and in 1823 built 42 booths for publicans. In 1853 they were to build[17] a Stand for Gentlemen, Trainers and Reporters; in 1867 they considerably altered the Grandstand.[18]

'Thomas', writes Hatfield, 'held that good company and good discourse are the very sinews of virtue, and as he admired civility and hated censure he eschewed the cantankerous individual'.[19] He was not averse to enjoying himself, and William Breach, an apothecary, submitted a bill[20] in 1766 which indicated that perhaps there had been a certain over-indulgence during the festive season. An extract of the account is as follows:—

Decr.	20	Bleeding	1	
		A Bals Mixture	1	8
		The same rep	1	8
	21	A Diapt [diaphretic] Mixture	1	10
	22	The same rep	1	10
	23	The same rep	1	10
	25	A dose of Physick	1	
	26	A Narcot Draught	1	
	27	A Mixture	1	10
	28	The same rep	1	10
	29	The Physick rep	1	

Thomas II married a lady by the name of Sarah from the village of Jump, near Barnsley. Her surname is unknown. The marriage is reputed to have taken place at Sheffield but research has failed so far to produce the marriage lines. Thomas II and Sarah had six children; Thomas Anelay III being the eldest. Twins were born in 1768;[21] John died in 1771, Maria in 1783. The inscription on the gravestone included the verse:

From this dull region here on earth
Our souls hath took their flight,
And gone a journey much more worth
To meet the Lord of Light.

Two of the three daughters married men of business; Sarah became the wife of John Mitchell, grocer of Bawtry; Harriet, of Richard Newsome, farmer of Armthorpe.[24]

The family home in 1781 according to Hatfield was in the 'back lane to St. Sepulchre's without North side'. This was renamed Golden Street in 1793 and later Factory Lane. Here Thomas II 'took pleasure in his extensive garden, surrounded by fruit trees, with hedgerows of Hawthorn'.[25] The street contained 15 houses of which he owned three. The short section of a later and wider Factory Lane remains at the rear of the theatre opposite Doncaster Railway Station. Presumably in 1798 when Catherine Thompson died the family moved into John Thompson's house in the Magdalens and lived here until at least 1871.[26]

Thomas II's death in 1798 came about as a result of catching a chill on a visit to Bawtry and he was buried next to the Thompson grave between the easternmost buttresses on the south side of the Parish Church.[27] In his will dated 1797[28] it is significant that he described himself as 'master bricklayer'. He was a craftsman who had learnt his trade, worked alongside his employees and then passed on his skill to another generation. He left his property to his wife Sarah for her lifetime. After her death in July 1812 there was an auction sale in December at the Woolpack Inn.[29] The sale notice included a house in Hallgate with 'convenient Out-offices, Garden and Appurtenances' in the tenancy of Dr. Webster; two newly built premises in Princes Street tenanted by Mr. Dey and Mr. Harrison, and three houses and 'appurtenances' in Factory Lane let to Thomas Greaves, Thomas Hotchin and William Stevenson. Thomas II's house in the Magdalens was bought by the Corporation[30] but re-acquired by the family in 1846[31] when the Corporation sold the property to raise money for the new Market Hall.[32]

Chapels on bridges were not uncommon in the Middle Ages; examples survive at Wakefield and

STATUARY AND STONE WORKS,
SILVER-SEREET, DONCASTER.

T. AND C. ANELEY

RESPECTFULLY inform their Friends and the Public, that they have entered to the Premises lately occupied by Messrs. F. CLAYTON AND SON, in SILVER-STREET, where the Business will be henceforth carried on in all its Branches, and where they hope, by strict attention to all Orders entrusted to their care, the employment of Skilful Workmen, combined with Reasonable Charges, to merit the support and patronage conferred on their predecessors.

Silver-street, Doncaster, May 27, 1846.

The EXECUTORS of the late Messrs. CLAYTON have Let the above Premises and Business to Messrs. ANELEY, and trust that they will receive the Patronage and support with which the business has been so long favoured.

8 (above) *An advertisement in the* Doncaster, Nottingham and Lincoln Gazette, *27 May, 1846.*

9 (right) *Thomas IV and Charles Anelay took over the premises in Silver Street (next to the shop on the left) which were to remain the offices and yard until 1918.*

10 (below) *The Market Hall, Doncaster, built by Thomas IV and Charles Anelay, 1847. Architect, John Butterfield, the Corporation Steward. The figure of Ceres on the centre pediment was carved by Anelay's foreman stonemason, Thomas Stenton.*

TRANSFER OF BUSINESS.

"T. AND C. ANELAY."

MESSRS. T. and C. ANELAY, of Doncaster, Stone Masons and Builders, beg to inform their Friends and Patrons that they have disposed of the whole of their business to WILLIAM ANELAY, and that he will receive all accounts and discharge all debts due and owing to and by the late Firm.

T. and C. ANELAY also beg to express their thanks for the liberal support which they have enjoyed during many years past, and their hope that their successor, WILLIAM ANELAY, will deserve and receive the same consideration and generous treatment.

Silver-street, Doncaster, ***16th February,*** **1875.**

11 (above left) *Hyde Park Cemetery, Doncaster, Gateway, Chapels and Lodge built by Thomas IV and Charles Anelay, 1855. Architect, R. J. Johnson of Newcastle. Brodsworth, Bath and Warmsworth stone were used for the archway.*

12 (above right) *The* Doncaster, Nottingham and Lincoln Gazette, *16 February, 1875 announces that William I has taken over the business from Thomas IV and Charles Anelay.*

13 (centre left) *Rossington Hall, Doncaster, built for James Brown by William I, 1882. Architect, unknown, ? S. S. Teulon.*

14 (below left) *St. Jude's, Hexthorpe, Doncaster, designed and paid for by Lord Grimthorpe and built by William I, 1893. It is of Balby bricks with Ancaster stone dressing and was Lord Grimthorpe's last church.*

Rotherham. There was one at Doncaster on the north road and the bridge was known therefore as St. Mary's Bridge. It was near here that Thomas II's brother, Charles, lived. He had come to Doncaster in 1775. A transcript of a lease of 21st December, 1783, refers to him as a blacksmith of St. Mary's Bridge End, Arksey, and his house is described as:

> Messuage and blacksmith's shop etc. at Bentley, in the parish of Arksay in the occupation of the said Charles Anelay, adjoining the river Dunn and St. Mary's Bridge End on the South and highway on the west, stable and beasthouse adjoining, piece of ground walled off from Catherine George's.[33]

Some outhouses behind the Three Horse Shoes may well be the buildings mentioned. In 1772 he was paid £1.1.8 for iron work to the water engine and for the shuttle at Crimpsall Wash.[34] Charles was also the innkeeper so the customer might be refreshed while the horse was shod.

A sale notice of 1795 advertises an auction at Charles' house.[35] He and members of his family are laid to rest on the right of the path from the south gateway of Arksey Churchyard. Hatfield lyrically records the impression of a visit there on 17th April, 1865.[36] Having failed to discover the 'orris-perfumed violets' in the hedgerow he rejoiced to find them close to Charles' grave. Of special interest to him was the grace of Charles' wife, Margaret, who died aged 37. He records the verse:

> Oh! cruel death that would no longer spare
> A loving wife and tender mother dear!
> Great is the loss to those who were left behind.
> But she, it's hoped, eternal bliss will find.

Their daughter, Margaret, was the wife of William Morley, a Doncaster merchant and the mother of Sir Isaac Morley, a worthy Victorian burgess of the town.[37]

CHAPTER IV

THOMAS ANELAY III (1764-1848)

WHEN THOMAS II died in 1798, his son, Thomas Anelay III succeeded to the firm. He had been apprenticed to his father in 1780 and the indenture is still preserved. Amongst the family papers there survives, too, an interesting relic of his youth in the shape of a leather bound almanac called *The British Merlin* of 1675. The full title is:

> Riders 1675. *Brittish Merlin*: Bedeckt with many delightful Varieties and useful Verities, fitting the Longitude and Latitude of all Observations of Principal Note, to this Year 1675, being the Third after Bissextile or Leap Year. With notes of Husbandry, Physick, Fairs, and Marts, Directions and Tables to all necessary Uses. Made and Compiled for the benefit of his Countrey, by Cardanus Riders. London: Printed by T.N. for the Company of Stationers. 1675.

On the interleaved pages are notes probably made by a man with the surname of Ricard [Riccard] the nephew of Richard Kellam. The Riccards of Heck had resided in the locality for many years. It is interesting to note that a family of this name lived in Doncaster where they were master carpenters, working at the Mansion House, doing the woodwork on Thomas II's gaol, and also at Cusworth Hall under George Platt.[1] Perhaps this little book was given to Thomas III by one of the Riccards as a plaything. Opposite one page are the following entries:

> 23^{d} ffebr '77 pd Ann or Servant an qtr
> of Mault for her 22s.
>
> 27th paid Mr. Geo Raisine Jun & Philipp
> Bassit for a yeares Rent to the
> Corparacion due at Michmas last before
> Wm. Perkyn the Mason.
> but I had not an Acquittance the summe
> at 4 li: 4 11d.
>
> 2^{d} ffebr 1677 Lent Buckley Wilsford
> wch I sent to him the Angell Tenn
> shillings by my unkle Kellam
>
> same day recd back 4s, 3s. 4d. whereof
> I paid for Wyne in his company wch
> Alderman Raisine is to bear it being
> about his businesse
>
> 16th March 1677 pd Wm Greene for the 4th
> quarter for building Shipps xviii d.
>
> 30th bought of Richd Newsor a load of Wheat
> called Purkey for wch I pd him x s.

Several of the names are those of senior Doncaster civic officials. George Raisine was an apothecary, Alderman in 1662[2] and Mayor in 1665;[3] Philip Bassett was Corporation Chamberlain in 1676[4] and at his death in 1679[5] was described as innholder; Buckley Wilsford was, no doubt, a relation of Theosebius James Buckley Wilsford who was removed from the Corporation in 1757 for obliterating an entry in the Courtiers' Book.[6] On another page there is an item about entering a bond with Mr. Daniel Hall for £50. He was Alderman in 1669[7] and Mayor in 1677[8]. A few years later he suspected that he would be displaced for 'knaveries and misdemeanours' and offered his resignation if the Corporation would allow him £30.[9] This proposal was refused. He sued the Corporation in 1679;[10] but it was not till 1685 that he surrendered his aldermanic office.[11] He was later restored and his tomb slab boldly declares the resting plate of 'Daniel Hall of Doncaster, Alderman'.[12] It is tempting to see an early reference to building in the entry for 16th March. 'Shipps', could be a Yorkshire Dales word for beast sheds; but this could equally mean ships.

Thomas III has signed the book several times and presented a riddle such as a joiner would set his apprentice:

'Make one pin or peg to fit the above three holes exactly.'

The Day Book of Thomas III is a large folio leatherbound volume of 325 pages written on both sides which describes in detail all the work carried out by the firm day by day from 1798-

1848. It is of considerable interest and importance for the architectural history of Doncaster and district. Taken with Thomas II's Day Book of 1772-1774 which also records the firm's wage bill, this collection contains much information about the economics and workings of a provincial building contractor during the period. For a detailed study see pages 31 to 60.

Hatfield says that Thomas III 'had a beneficial arrangement with the Doncaster Corporation; for it was evident that he had all their work'.[13] And, in fact, the Corporation entries are by far in the majority: but other firms were also employed and the Anelay share diminishes over the years. In 1801 there are details of the considerable amount of work under 'Mr. Lindley's Day Account for repairing the Mansion House'. This is probably the addition of the attic storey built that year. £729.8.10 was paid to 'T. Ainley (builder) for measurework and day work'.[14]

Public works included the building of weirs and bridges for the Commissioners of Carr Drainage in 1818 and 1820.[15] There were alterations to substantial property: coach houses for Mr. Ramsay, the licensee of the Corporation Arms, Rossington Bridge 1806, town house and stables for Mr. Francis Hill in Car [*sic*] Lane 1817, and considerable additions to the High Street house of Mr. John Henry Maw 1801-3. A new house for Capt. Ramsden of Hexthorpe in 1842 used 49,080 clamp bricks. Small jobs were carried out on the larger houses in the neighbourhood, Thomas III apparently having a special line in doing the complicated flue work for the Count Rumford grates (see page 49). He worked for Lord Stourton at Stapleton in 1799 and for the Hon. Frederick Lumley at Tickhill in 1802.

From the study of the Day Books below it would appear that originally the firm specialised in brickwork, doing a little colour washing, slating and tiling, stonework and carpentry on occasions. About 1842 however, the firm apparently began stonemasonry in a serious way. There is a considerable amount in the accounts for Captain Ramsden's house mentioned above and in 1843 the firm was awarded the contract for the completion of the restoration of the East end of the Parish Church of St. George's.[16] As will be related below, Joseph Lockwood, junior, the stonemason, whose family friendship with the Anelays had lasted two generations, died in 1842. Lockwood's business was carried on by Messrs F. Clayton until 1846 when Anelays entered into the Statuary and Stonework premises in Silver Street and Dockin Hill Yard.[17] It would appear that the Anelays may have thought that Lockwood's death had provided an opening as they started to carry out stonemason's work in 1842. Perhaps they took on some of Lockwood's stonemasons.

As for ecclesiastical work, the first pages of the Day Book record in detail the conversion of a barn into the Quaker Meeting House in West Laithgate, Doncaster, in 1798. One end of the barn was made into a caretaker's cottage: the main part a hall for 200 people. There was also a stable. The building later became a centre for Adult Education at the start of that movement, with working-men's classes beginning at 8.00 a.m. on Sunday morning.[18] There is also an account for minor alterations to the 'Calvernist Chappel' in 1798 and 1800 which was the predecessor of the present Congregational Church in Hallgate. Apart from work on the Parish Church there is little mention of Anglican Churches, though at Rossington the school house was converted to a rectory and they repaired parochial property for the churchwardens of Doncaster. Local incumbents, too, were amongst the firm's clients.

The collection of payments due to Thomas III were at times difficult and he writes on one, perhaps a poor widow, 'Never to be received'; of another, Mr. Pearson's 'I do not expect this acct. to be any more settled than this'. Amongst the scraps of paper in the Day Book is the following recipe:

1 7½ pound of Quick lime to 12 Quarts of
Watwe Boiled Down to 6 Quarts
2 Pounds of Burdock Root 2 large Handfulls of
Bugbear 1 pound of the yellow Dock Root
2 Handfulls of the Heartsease 2 Handfulls of
Colts Foot 1 Handfull of Robbin run in the Hedge
1 Handfull of Camomile All Boiled Together and
Strained ¼ pound of Epsom Salts.

The purpose of this almost magical concoction is not known.

Thomas III married Hannah Doncaster at the Parish Church in 1799: the ceremony was witnessed by Richard and Mary Anelay, his cousins.[19] Two of the sons of this union, Thomas IV and Charles, were later to enter into a business partnership. Thomas III, whom William Sheardown in his pamphlet *Moat and Mound* names as a 'respectable and intelligent mason',[20] died tragically in 1848 as the result of having consumed a glass of vitriol and water which he had mistaken for beer.[21] He was buried in the family grave in Doncaster churchyard.[22]

CHAPTER V

THOMAS ANELAY IV (1805-1886) and CHARLES ANELAY (1808-1896)

DURING Thomas Anelay III's lifetime a trend indicating the firm's direction began to be discernible. There was and would continue to be some work from the corporation, the racecourse, houses large and small and churches; but during the partnership of Thomas IV and Charles the most marked development was the work undertaken for architects of not only local, but national, repute. They repaired Hatfield Church, Yorkshire, in 1846[1], and an early major building was the Market Hall, 1847[2] (architect, John Butterfield, the Corporation Steward). The Corporation invited tenders from different craftsmen as opposed to the employment of a single contractor and estimates were submitted for Ironwork, Plumbers', Slaters', Joiners' and Plasterers' work and 'The Masonry work etc.' Anelay's estimate for £2,934 was a greater amount than the rest of the skills put together.

The Magdalens, Doncaster's market place since 1575 where the family house stood, contained a group of buildings built into the remains of the medieval church of St. Mary Magdalene.[3] As the additions were removed the skeleton of the church was revealed. A controversy then ensued between those who would have the ruins preserved and those who were for removing them. Led by the vicar, a memorial was sent to the Corporation; referring to the fact that

> the remains of a building once consecrated to the worship of God and interment of the Christian dead claim a tender and respectful treatment.[4]

A plea was put forward that the ruins be preserved as

> a specimen of Ecclesiastical Architecture as a time honoured memorial of our holy faith as an illustration of the history of this ancient town.[5]

The Corporation was confronted with a difficulty as this was the most valuable site in the town. The Memorial was duly received and a reply issued on 10th February, 1847. The Town Council would have been happy to let the ruins remain as at St. Mary's, York, but in view of the position of the site they felt sure that

> public utility would not be promoted by a compliance with the wishes expressed in the memorial.[6]

So the Corporation offered the ruins to the memorialists. Some pieces found their way to the garden of Beckett's Bank in High Street; others to the gardens of the principal citizens of the town. The figure of Justice which crowned the pediment was taken down and stored in Anelay's yard in Silver Street with a view to its re-erection on the new Guildhall when it should be built. Thus, Thomas Anelay IV and Charles for the first time came into close contact with medieval masonry; they had their eyes opened to the controversy that can surround the questions of Conservation. During the excavations of the site, they discovered a Bronze Age cinerary urn which was later presented to the Doncaster Museum.[7]

Municipal and other large contracts included Doncaster Cemetery Chapel, Lodge and Gate (R. J. Johnson of Newcastle on Tyne), 1855[8]; alterations to Doncaster Vicarage, 1860;[9] improvements to the racecourse Great Stand, 1861;[10] a new wing to Badsworth Hall, 1861;[11] the Infirmary[12] (now the Education Offices) in Wood Street (Brundall and Arnold), 1865; Great Northern Railway Gas Works,[13] 1866; refacing Becketts Bank,[14] High Street (B. S. Brundell), 1868. The firm built Barnburgh Rectory[15] (Thomas C. Hine, Nottingham), 1850; a stone gabled house in Regent Square, 1868.[16] The first complete church they built was St. John's, Doncaster, to the design of Thomas Penrice, of Clapham, 1870[17]. They had added a new vestry at St. John's mother church, Christ Church[18] in 1856, and carried out the restoration of an Anelay connected church, Arksey in 1868-70.[19] In 1849 they restored Kirk Bramwith Church.[20] On one occasion a grateful client expressed his appreciation by entertaining all the members of the firm. On 15th October, 1855, the local Press[21] reported that

> there had been a dinner the previous Saturday when about forty sat down to an excellent dinner at the house of Mr. Cox of the Dolphin Inn, given by T. Walker Esqre where they have been erecting extensive premises . . .

From Directories it would appear that the firm used the Magdalens where they lived as an office and yard until the leases of Silver Street and Dockin Hill with its wharfing facilities besides the River Don were acquired in 1846 from Mr. F. Clayton (who worked the Roche Abbey Quarries); he in turn had bought it from Mr. Joseph Lockwood (1758-1837). Lockwood had built three bridges in 1795, and Hatfield[22] maintained that these were the foundation of his future success in building; he was a trustee of Thomas II's will and had opened his stoneyard in Silver Street in 1790. He took his part in civic affairs and was an alderman, being Mayor in 1823 and 1832. His son, Joseph Lockwood (1785-1842), was a sculptor, and his grandson, Henry Francis Lockwood, an architect[23] (1811-1878). The latter did much work in Bradford, and in London he designed the City Temple (1873-4) and the Civil Service Stores (1876-7).

Thomas VII used to tell how the clerk recalled that a frequent visitor to the stoneyard was the murderer Charley Peace. It was no coincidence perhaps that he was defended at his trial by the great grandson of the original owner of the Silver Street yard, Frank Lockwood—later to be knighted as Attorney General.[24] Born in 1832, Peace quarrelled with his next-door neighbour in Banner Cross Terrace, Sheffield, in 1877, and killed him. He sought refuge in London where he lived a double life, He rented a basement flat and would play his violin at respectable *soirées* upstairs and then later emerge in disguise with his violin case packed with housebreaking instruments. He was finally caught, tried and sentenced to be hanged. 'He was calm during his pre-execution breakfast, delivering a little homily to the warders on a Christian state of grace, but a flash of the true rascally Charley came when he glowered down at his plate and observed: "This is bloody rotten bacon".'[25] Perhaps it was the memory of what he had seen in Anelay's Yard that enabled him to fabricate out of old memorial cards and mounts a fantastic design for a mausoleum (never executed) which was to be finished in marble and gold. The model is now at Madame Tussauds.[26]

From the Census records of 1841,[27] 1851[28] and 1861,[29] it is possible to compare the respective households of Thomas IV and his brother and partner Charles Augustus, their status and the expansion of the business. There were three generations of Thomas Ainleys [*sic*] living in Scott Lane, aged 75, 30 and 7 years' old respectively, Thomas III, IV and V. Also resident were Thomas IV's sister Harriet: his aunt Sarah Doncaster, and his two daughters Marion (aged 4) and Maria (aged 1); and two apprentices, William Thompson and John Birley. In East Laithgate lived Charles and his wife Elizabeth and their three children, Charles, Ann and Harriet. Ten years later at the Market Place, Thomas IV no longer described himself 'brick-layer' as his father did but 'builder'. His immediate family had been increased by the birth of William I, his aunt Sarah was still resident and also his sister, Harriet, now described as school mistress. Times were evidently more prosperous as he was able to employ a servant, Ruth Wilkes. In 1861, at '3 Magdalens', there was not only a servant but a governess, Mary Bond, born at Goole, who helped in a school now run by Harriet. In this census return Thomas V was described as 'Mason', presumably having served a normal apprenticeship before his excursion into architecture.

Meanwhile, Charles Anelay's household having moved to Silver Street by 1851, in 1861 remained much the same. His son, Charles, in 1861 was described as a mason, and in 1867 was clerk of works to the Midland Railway, perhaps through the good offices of his cousin Sir Isaac Morley, a director of the Company. The establishment of 13 men and four boys of 1851 had risen to 41 masons, 11 men and one apprentice in 1861. Charles Senr. retired in 1875 to Stainforth Hall, described as 'a handsome little hall situate near the canal navigation and the river, now turned into a farmhouse with extensive outbuildings'.[30]

It is strange that there is so little information about the personality of Thomas IV. Perhaps some impression may be gained from the fuller account of his sons, Thomas V and William I. Both sons took an interest in building. The elder, Thomas V (1833-1917), though he worked for some years as a qualified mason was more interested apparently in architecture and did not remain with the firm; he studied in the London office of the Victorian church architect, Samuel Sanders Teulon (1818-1873).[31] Teulon's churches are somewhat massive by present-day standards but are more and more appreciated by the discerning. His work evoked considerable appreciation from the Ecclesiological Society, particularly St. Andrew's, Lambeth, 1856.

In 1862 John Butterfield expressed his wish to resign his post as Doncaster Corporation Steward and Surveyor. He was prevailed upon to stay upon the provision of an assistant. Originally, the Corporation's intention had been to appoint an assistant themselves but out of respect for Butterfield they had agreed to leave the choice to his hands, and he chose Thomas V.[32] He succeeded his senior in the following year, later to become extolled as 'the zealous and active Borough Surveyor'.[33] In due course he also became Steward, retiring from both offices in 1888.

To honour their highly respected fellow-citizen, the Corporation appointed him Borough Remembrancer in 1898, a post that was not revived until 1944.

In a series of articles on Worthies from the past in the *Doncaster Gazette* is an appreciation of his very special qualities.

He was:

> a gentle-mannered pleasant voiced man. He affected the side whiskers of the early Victorian era, and he wore a collar and stock of an even earlier period . . .
>
> He could talk about Royal visits, grants and deeds: the rights and privileges of the burgesses: the history of our churches, markets and civic buildings; the growth of our institutions, the establishment of our schools, the rights of freemen—indeed tap him on any question concerning the past and there welled out of him a stream of information.[34]

His obituary relates:

> In temper he was courteous and kind, gave freely to the needy and to all good causes. His clean picturesque figure, so conspicuous in our streets these many years has disappeared from our midst and his pleasant and benevolent countenance and his cordial greetings have become a memory.[35]

He was a devoted Public Servant who took pride in the friendship of his brother-in-law Alderman Henry Wainwright who was Mayor of Doncaster during the Jubilee celebrations of 1887. He must have been proud, too, of his sister Marion, the First Lady for that Royal Year.

Thomas V was a keen churchman and six times churchwarden of the Parish Church. A methodical man he would walk every Sunday the same route to church, via Silver Street, Bower's Fold and the Market Place. Amongst his many friends was Alderman Bentley, the brother of John Bentley who, after spending his childhood in Doncaster, was later to be the architect of the Roman Catholic Cathedral in Westminster. There could perhaps be no finer tribute to this distinguished public figure than this:

> Mr. Anelay was the last Borough Steward and Remembrancer; there was nobody to put in his place. And so this ancient office was extinguished and the link with the medieval past was snapped for ever. It is not likely there will be another Thomas Anelay.[35]

CHAPTER VI

WILLIAM ANELAY I (1841-1918) and THOMAS ANELAY VI (1866-1942)

OWING to the advancing years of the two partners in the firm, it was imperative that measures be taken for its continuance. These were revealed in the local Press on 19th February, 1875[1] where an advertisement stated that Messrs Thomas and Charles Anelay had disposed of their business to William Anelay, and after thanking their past customers they hoped that their successor would deserve the same kind and considerate treatment. What financial arrangements were made do not appear, but it is clear that the transfer of the business caused a rift between the two cousins,

William I and Charles junr., for in the following issue of the *Gazette*[2] an announcement appeared intimating that owing to his father and uncle 'having given up business', he (Charles) had taken premises in High Street Buildings where he intended to carry on the building business in all its branches. This is reflected in Kelly's Post Office Directory of 1877 where William occurs as a 'Stone Mason' at 31 Silver Street and Dockin Hill, while Charles is entered as a 'Builder' at 1 High Street Buildings. In 1883 Charles moved to Beeston, Notts., and his son and grandson became builders in Coventry.

The first large building William I constructed was Rossington Hall, completed in 1882 for James Brown, M.P. In 1889 he built the Doncaster Free Library. This project had been a matter for competition and the winning architect was H. A. Cheers of Twickenham. In 1894 he built St. Jude's Church at Hexthorpe which was Lord Grimthorpe's last work.[3] The original designs for this were drawn up by him; but the work was completed under Herbert Athron. Anelays had had early dealings with Lord Grimthorpe's family; as Edmund Beckett he had been one of those largely instrumental in bringing the Great Northern Railway Locomotive, Carriage & Wagon Works to Doncaster, and in 1853 the firm had built their Gas Works, in 1868 they had refaced Beckett's Bank in Doncaster. Lord Grimthorpe was an amateur architect of peculiarly forceful character. Perhaps his special claim to fame is his design of the clock in Big Ben; but his restoration of St. Alban's Abbey is a testimony of an ability to recreate a medieval building, although we find it hard to think of this church as anything other than a Victorian piece of architecture. He was Chancellor of the Diocese of York and had little patience with the antiquarian. When his attention was drawn to the fact that Norman and Saxon fonts from churches in the diocese were being put up for sale, he replied on a post card that 'he had no more power or wish to prevent a man selling his old church fonts than his chinese geese or three tons of manure'.[4] Anelays were later to work on Grimthorpe property at Easthorpe Hall and Westow.

Church restoration continued to occupy the firm. In 1885-6 they restored St. Leonard's Church,[5] Wollaton, and in 1890-1, Marr Church[6], the architect being on each occasion Charles Hodgson Fowler. There is a tradition that William I carried out the restoration of Hickleton Church under the supervision of G. F. Bodley in 1886[7] for the first Viscount Halifax. A later generation was to work for the third Viscount at Garrowby in 1959, for the fourth in 1960 and in 1974. The firm also made the memorial to the third Viscount in the churchyard at Kirby Underdale in 1962. In 1893 they built a private chapel adjoining Hesley Hall from designs by Henry Christian of London.

> An arrangement has been made by the Bishops of Southwell and Rochester whereby the chapel will be served by one of the large staff of St. John the Divine, Kennington, in order to afford the clergy of this South London parish, country air, whilst material aid will be given by Mr. Whitaker to St. John's.[8]

William I restored the church at Cantley in 1894. This was refurnished in the distinctive ritualistic style by Mr. (later Sir) Ninian Comper. The newspaper reporter, of a low church persuasion no doubt, described the re-opening service which was conducted 'according to the rites and usages of the extreme Ritualistic party'—and there were all the usual accessories—vestments, lights, incense, sanctus bells, etc. He comments on the use of candles:

> In olden times when the windows of the churches were narrow and not so numerous as present, artificial light was a necessity and candles had to be introduced, but nowadays candles are placed in all parts of the chancel without rhyme or reason.

He refers to the 'gregorian' music and 'how the first two lines of *Salve Festa Dies* were sung by the clergy (indifferently sung, too)':

> On reaching the chancel the clergy sang the office—'O how dreadful is this place'—and meanwhile a considerable amount of incense was used until the chancel was enveloped in a mist as dense as a fog on a raw November morning.[9]

In 1906 the church was gutted by fire and another restoration was necessary. Again the architect was Mr. Comper.

The intervening years had increased the understanding of High Church ritual. The ceremonies were now taken for granted and a less florid report appeared in the newspaper[10] which provides, nonetheless, a delightful vignette of a High Church celebration in the early 1900's. The Preacher at the High Mass was Father Stanton, the famous Curate of St. Alban's, Holborn. There was dancing round the Maypole on the Green afterwards and the Pom Pom Pierrots entertained the parishioners at night.

William I married Miss Clara Milan at Doncaster Parish Church in 1865. They had four children: two sons and two daughters, Jessie, and Evaline who died in infancy. The elder son Thomas VI succeeded to the firm; the younger, Walter, died by his own hand at the age of 38.[11] It has already been mentioned that the families of Anelay and Bentley were friendly. Walter and John Bentley the architect of Westminster Cathedral knew each other and it would seem that John used to spend some time in the Anelay workshops. His biographer, De L'Hôpital, says that he had been accustomed to free entry into the carvers' and joiners' workshops in the town.[12]

> The hours spent in the masons' yard and at the joiners' bench were soon the happiest in his life, for the men took pleasure and pride in initiating the intelligent boy into the mysteries of their crafts.[13]

Perhaps the firm was employed in 1890 for overhauling and decorating the Bentley home or they may have worked together at Carlton Towers, near Selby, between 1875 and 1891.

William I cut a similar figure to his brother Thomas V, the Borough Remembrancer.

> Mr. Anelay never took part in any public affairs and held "the faith of our fathers" as represented by the Church of England with true Christian modesty, and his principles as a supporter of the Conservative party and Constitution without fuss.[14]

His obituary continues:

> The two men [William I and Thomas V] were the conspicuous representatives in their day of a family of which old Doncaster is proud—happily not the last and well favoured. The picturesque brothers Anelay will not soon fade from the public memory.[15]

Reference has already been given to the way that the firm developed relationships with particular architects. In 1898 their collaboration with Mr. Walter H. Brierley on the building of the York City and County Banking Company—now the Midland Bank at the corner of High Street and Baxtergate, Doncaster—was to have a considerable effect upon the future of the firm. Walter Brierley was a partner in the firm of architects founded by John Carr in York in the 18th century.[16] The friendship between William I's son, Thomas VI, and Brierley deepened and he persuaded Thomas to open an office in York. The Doncaster connection continued with the office in Silver Street remaining open until 1918, and the yard at Dockin Hill until 1939. It is, in fact, from the Brierley office records in York that the work of the firm under Thomas VI is to be traced.

In Thomas Anelay VI's time came the building of a new wing for the York School for the Blind, at the King's Manor, also the construction of several large private houses, among them Grimston Court with its lodge and Thorpe Underwood 'in tasteful neo-Tudor' (N. Pevsner).[17] Another notable building was Brierley's own house, Bishopsbarns, St. George's Place, York, which evoked this letter of praise from the architect: 'Everyone admires the work and I think that it will hold its own as an example of what could be done in 1905, in a plain and simple way'.[18] There was also a great deal of conservation work including repairs both to secular buildings such as Middleham and Gilling Castles as well as Yorkshire churches like Hutton Buscel, South Kirkby and Foston. Thomas VI also built churches—at Glasshoughton (1900) and St. Luke's, York (1910), to Brierley's design. Anelay interest in the turf was to be found in a contract for the racecourse at York in 1909.

In 1920 the Diocese of York Advisory Committee on Faculties was set up by Archbishop Lang, and among its first members were Walter H. Brierley and John Bilson, both of whom were skilled in the conservation of ancient buildings.[19] Anelays worked under the latter on several commissions, among them the restoration of Escrick Church after the fire of 1923. Mr. Bilson testified to the excellence of the builder's work as follows:

> Mr. Thomas Anelay of York has carried out for me most satisfactorily many works of repair on ancient churches extending over many years and I can recommend him most strongly. Indeed I do not know any builder in the North of England who is as capable, trustworthy and reliable for such work requiring special care and attention.[20]

In 1970, Thomas VI's grandson, William Peter Anelay, was to become a member of the same committee, now called the York Diocesan Advisory Committee for the Care of Churches: and in 1974 became Chairman.

Such indeed was the firm's accumulation of skill that it is not surprising to hear that after they had worked for Charles Peers, repairing the Marygate Wall of St. Mary's Abbey, York, many of the repairing techniques became the accepted practice of the Ancient Monuments Department of the Office of Works.

Thomas VI held responsibility for the firm during and after the Great War. In 1916 they received the contract for building workshops for Messrs. T. Cooke & Sons on Bishophill in York. Messrs. Cooke manufactured optical instruments essential for the war effort. Important commissions[21] included blocks of buildings for St. John's College, York, and for the Community of the Resurrection at Mirfield; the restoration of Middlethorpe Hall and Nunnington Hall; and the building of Goddards on Tadcaster Road, York. Perhaps the most important work during Thomas VI's time, at least from an archaeological standpoint, was the restoration of the Merchant Adventurers' Hall, York, a building whose true importance has yet to be appreciated. Brierley had complete confidence in his builder and when asked by Cecil Leckenby about the wall at the north-east corner he wrote:

> As you know in old work I always prefer an old wall repaired if it can be well and substantially done but . . . I leave it to Anelay and you to do what is necessary.[22]

In 1918, Thomas VI came to live at 46 Bootham in the parish of St. Olave's, York, until he sold the premises to James H. Rutherford, Brierley's partner. The distinctive features of Thomas VI are perpetuated wittily by the York carver, Wilfrid Milburn, in the bearded figure smoking a pipe, which is carved in the pediment over the main door of Thomas' Hotel in Museum Street, York. Thomas VI was a man of generosity, a keen churchman and a churchwarden for a time at the parish church in Marygate.

CHAPTER VII

WILLIAM ANELAY LIMITED

IN 1927 Thomas VI retired and his place was taken by William II, the firm becoming a limited liability company the same year. During the slump, Anelays were fortunate in being employed by Commander Clare Vyner on extensive repairs and alterations at Fountains Hall and Abbey, Studley Royal and other estate buildings. Later the firm resumed the same sort of building work as in previous generations. A notable sign of the times is the absence from any list of the building of a large country house *de novo*. There were plenty of alterations—West Heslerton Hall 1934, Escrick Park 1948, Huttons Ambo Hall 1950 and Everingham Park 1963, as the larger houses were adapted for more convenient use. Churches included the Roman Catholic Chapel at Catterick Camp 1929, the Church of All Saints, Dormanstown, Redcar 1932, and the Church/Hall of James the Deacon, York 1955. Restoration was carried out on Filey Church 1929 and some conservation work at Castle Howard 1932 and 1940, York Assembly Rooms 1940-1, and the Judges' Lodging, York 1953.

The Second World War interrupted the firm's normal working. They were engaged on carrying out War Department Term Contracts in York and the East Riding. After air raids, war damage repairs were carried out on behalf of York Corporation. When the war ended and housing was the main priority, four houses were built on the Water Lane estate, Clifton, York 1945-6, eight houses at Great Habton and 10 at Hovingham for the Malton Rural District Council 1946-7. A large contract was the building of extensions to Ripon Diocesan Training College 1951-3.

William II was joined by his son William Peter in 1945 who assumed the position of Managing Director in 1965, William II remaining as Consultant. Over the past three generations of the Anelay family the firm has continued to carry out its specialist work. To William II must go the credit for adapting an old established family company to the needs of the present day and this continuing process is being carried on by his son, joined by Mr. Peter M. Johnston in 1968 who was appointed a director in 1970. There are few large houses in the York area which have not at some time displayed the distinctive Anelay sign-board, which along with the letterhead motif were designed by John Snowden, a student at the York College of Art in 1962. The institution of grants by the Historic Buildings Council has given the firm work at Belton House, Grantham (Architect F. F. Johnson) 1967-72; Gawthorpe Hall, Lancashire (F. F. Johnson) 1968-9; Newby Hall, Ripon (Ward, Ruddick and Ward) between 1957 and 1972.

A glance at some of the recent work undertaken by the firm is indicative that their special interests are the same as they have been throughout the development of the business: for example, restorations at York Castle Museum 1971, Bramham Park 1967-72, and Holy Trinity Church Goodramgate, 1973-4, York County Savings Bank 1970, conservation of the ruins of Kirkstall Abbey 1971, and alterations to Bishopthorpe Palace 1971. In 1972, Anelays restored Bishop Cosin's Library at Durham, and the exterior of the Tunstall Chapel in Durham Castle. It is a happy coincidence that Anelays returned to work if not for the Corporation of Doncaster, then for its Rural District Council, in repairs to Cusworth Hall in 1970-4.

Conservation has long been an Anelay speciality many years before that word became fashionable. Among the most recent examples is the Royal Coat of Arms over the main gate of King's Manor, York. It has been carved, repaired and conserved to be as splendid an entrance as those of St. John's or Christ's College, Cambridge. In progress are two projects involving the conservation of derelict houses of historic and architectural importance, acquired by the firm. Manor Farm, Gristhorpe, near Filey, is a limestone Wold farmhouse dating from 1747 which has a stone plaque over the doorway from a previous house dated 1665. Temple Farm, Temple Hirst, on the road between Carlton and Chapel Haddlesey, near Selby, incorporates the remains of a Preceptory of the Knights Templar.[1] Although extensive reconstruction has taken place in the 17th and 18th centuries, a reset Norman doorway, medieval tower and an old wood spiral staircase survive.

Reference has already been made to the repair of Cantley Church, which involved the making of church fittings and furniture. Since that time the firm has enjoyed the privilege of making first-class furniture for specific churches to architects' designs: at St. Mary's Beverley, Armagh, Llandaff and Durham Cathedrals for Mr. George G. Pace, and at Ripon Cathedral for Messrs. Ward, Ruddick and Ward. A further activity in which Anelays have been engaged for many years but which has considerably increased since 1960 is the treatment of woodworm infestation and outbreaks of dry-rot. Work of this type requires knowledge and experience coupled with practical skill and thoroughness. Many historic buildings in Yorkshire and beyond have been treated by the firm. A close association involving training and information, has been built up with Messrs. Richardson and Starling Ltd. of Winchester, the manufacturers of Wykamol and one of the leading experts in the field.

Awareness of the need for economy and efficiency in a technological age led to the adoption in 1971 of an integrated accounting system installed by the Building Advisory Service which amalgamated the costing and financial accounts, and in 1974 a new planning system was also installed by the same organisation. From 1918, when the Coney Street office opened, the firm's depot in York was at Skeldergate. In 1939 Anelays acquired the War Memorial Hall, Murton Lane, Osbaldwick.[2] The reason for this particular property being put on the market may have derived from the inaccessibility of the site to the two communities of Osbaldwick and Murton which it was to serve or the current war scare. It was only after the war was over that the housing expansion of this area really took place though the site as a social centre is still too remote even now. In 1963 the Coney Street office was closed and the administration of the firm integrated on the Osbaldwick site.

Any organisation is as good as the people who constitute it and no survey of the firm would be complete without mention of the many employees who played such an important role in its history. To encourage and look after good craftsmen and workers is axiomatic with the firm. In an industry where the custom has too often been a policy of 'hire and fire', Anelays have always tried to keep good men and to encourage sons to follow fathers. In the first list of Thomas Anelay II's employees in 1772 there was not only 'Self' and 'Lad' but also 'Slack' and 'Son', and 'C. Scales' and 'Son'. Members of the Slack family had a continuous connection with the firm for nearly 150 years. John Slack was an employee in 1861 and Jack remained till his retirement in 1919. Since 1974 Anelays have set up a Pension and Assurance Scheme which provides pensions on retirement, for widows, and life insurance cover.

An employee of special note was Thomas Stenton, a stonemason, sculptor and carver, and foreman for Thomas IV and Charles Anelay, who carved the figure of Ceres on the new Doncaster

Market Hall in 1847 together with the monument to Frances Bulmer[3] in the Parish Church of Doncaster. He was still foreman for the repair of Arksey Church in 1868-9[4] for Sir George Gilbert Scott. Two worthy present-day successors are Frederick Staples who has been the firm's General Foreman since 1951 and in charge of all the principal contracts during this time, and Mr. Lyn Letby the present foreman stonemason who has done carving at Belton House, Grantham, Newby Hall and the County Assize Courts, York.

In 1897 a photograph was taken of some of the firm's members. There were three stonemasons, two scaffolders, and an engine driver. Joinery was a flourishing department; there were six men and two apprentices and the foreman is notable for his white apron and bowler hat.[5] On the photograph there is a young man, Jack Chantrey, who started for the firm when he left school at 13, and worked until he retired in his late 70's in 1945. He remained an expert scaffolder and stone sawyer with unrivalled skill in handling heavy or bulky materials.

The present joiners' shop foreman (since 1958) is Mr. Charles F. Oldfield who was apprenticed with the firm as a joiner when he left school in 1944. His father was disabled and worked as a gardener; but his uncle, Stanley Oldfield, also a joiner, worked for Anelays from 1928-40. So did his grandfather, George, a mason, at such jobs as Keldy Castle and Nunnington Hall, in the period 1910-30. George's two brothers, Jim, a joiner, and Dickinson, a mason, both worked at Keldy Castle and Jim continued at Nunnington and Doncaster. Dickinson's son, Walter, was a highly skilled bricklayer and worked for the firm from the 1920's until his death in 1950. He was a most capable general foreman for much of the period. In all six first-class craftsmen from the Oldfield family worked for Anelays.

There has always been a happy relationshop between the family and the workmen, as is to be shown by the dinner given by William I in 1887 on the occasion of his son Thomas VI's coming of age. Between 50 and 60 were entertained to an 'excellent repast' at the Woolpack Hotel. Suitable toasts were drunk, after which Mr. Marsh, the foreman, was elected to the chair. In the course of the evening Mr. W. Turner who had been in the employ of the firm for 36 years presented to 'Mr. Tom Anelay a handsome marble timepiece on behalf of the employees'. This clock was for many years in the family's possession.[5]

This brief historical sketch has attempted to tell the story of the Anelay family; to describe the development of the firm, and to outline the main types of work the firm has undertaken in each generation. From small beginnings in the 18th century this business has adapted its skills and organisation to the needs of the times. Often the growth of a business tends to render impersonal its dealings with the client but Anelays continue to provide the family touch. Maybe the days are over of providing large municipal buildings such as the Doncaster Market Hall and the construction of large country houses within a confined radius. Now there is an increasing demand for the adaptation and repair of ancient buildings and churches all over Yorkshire, the Midlands and the North. In all Anelays' work, the high standard of their sound craftsmenship persists; as does also that special relationship between client, architect and contractor. Today, when there is a real danger of losing inherited skills of generations of craftsmen, Anelays carry on in the two-and-a-quarter-century old tradition of John Thompson, 'a clever and skilful masterbuilder'.

NOTES

Chapter I John Thompson

1 Miscellaneous Bills, Receipts, etc. Parcel 26, Doncaster Corporation Archives
2 Doncaster Marriage Registers, Vol. III, p. 144
3 C. W. Hatfield, *Historical Notices of Doncaster*, Vol. II, p. 290
4 John Wesley's Journal (Everyman Edition), Vol. I, p. 527
5 Hatfield, Vol. I, p. 30
6 J. Tomlinson, *Doncaster from the Roman Occupation to the Present Time*, p. 222
7 Courtiers Book, Vol. III, p. 806/788
8 Hatfield, Vol. II, p. 290
9 Tomlinson, p. 219
10 Hatfield, Vol. II, p. 290

Chapter II Anelay Origins

1 George Redmonds: *The Ainleys of Elland in The West Riding of the County of Yorkshire*
2 Op. cit., p. 1
3 Op. cit., p. 2
4 Op. cit., p. 3
5 Op. cit., p. 5
6 Op. cit., p. 6
7 Op. cit., p. 5
8 Register of the Freemen of the City of York, Vol. I, p. 269
9 Register of the Freemen of the City of York, Vol. I, p. 279
10 Parish Register of St. Martin Coney Street, York. YPRS, 1, 2, 3, 72, 74, 75
11 Parish Register of St. Olave's, York. YPRS LXXII, 52, 55, 62, 108
12 Register of the Freemen of the City of York, Vol. II, p. 37
13 Redmonds, op. cit., p. 7
14 J. W. Clay, Registers of Wath upon Dearne
15 C. E. Whiteing, Registers of Hooton Pagnell, 1538-1812
16 C. E. Whiteing, Registers of Brodworth, 1538-1812
17 All references are from the Snaith Parish Registers unless otherwise stated
18 G. Lawton, *Collectio Rerum Ecclesiasticarum de Dioecesi Eboracensi*, Vol. I, p. 155
19 Register of Carlton juxta Snaith
20 Hatfield, Vol. I, p. 30
21 *Doncaster Gazette*, 17th June 1870
22 Hatfield, Vol. II, p. 328
23 Hatfield, Vol. II, p. 432, 3
24 Day Book, October 1773
25 Hatfield, Vol. I, p. 30
26 Mentioned in the will of Richard Anelay of Stainforth, 1782
27 Hatfield, Vol. I, p. 30

Chapter III Thomas Anelay II

1 Day Book
2 Miller, *Doncaster*, p. 184, gives the wrong date 1773. Day Book
3 Credit Book 1772/3 to 1783, p. 107 Doncaster Central Library
4 Records of Doncaster, Vol. IV, p. 261
5 Day Book
6 Day Book
7 Day Book
8 Cash Book 1783-1797, p. 58. Doncaster Central Library
9 Hatfield, Vol. I, p. 31
10 Day Book
11 Day Book
12 Doncaster Corporation Credit Book 1772/3 to 1783. 'To My Note for £100 dat 19 June 1779'
13 Hatfield, Vol. I, p. 35
14 Ledger 1797-1808, p. 104. Doncaster Central Library
15 Hatfield, Vol. I, p. 31
16 J. Tomlinson, *Doncaster*, p. 207
17 *Doncaster, Nottingham and Lincoln Gazette*, 3rd September 1853
18 Pamphlets, Vol. II, p. 580. Wm. Sheardown, Doncaster Central Library

19 Hatfield, Vol. I, p. 31
20 Original in the family's possession, quoted in Hatfield, Vol. I, p. 31
21 Hatfield, Vol. I, p. 440
22 Hatfield, Vol. I, p. 427
23 Hatfield, Vol. I, p. 468
24 Hatfield, Vol. I, p. 468
25 Hatfield, Vol. II, p. 525
26 J. A. Harrison, *Private Schools in Doncaster in the Nineteenth Century*, p. 18
27 Hatfield, Vol. I, p. 31
28 Borthwick Institute, March, 1798
29 Sale Notice, Anelay MSS
30 W. Hurst's annotated copy of Edward Miller's *History and Antiquities of Doncaster*, p. 188-9
31 *Doncaster, Nottingham and Lincoln Gazette*, 24th July 1846
32 Hatfield, Vol. I, p. 414
33 Sykes Collection. Notes of Original Deeds, Doncaster Central Library
34 Doncaster Cash Book, 1772/83, p. 95
35 *Doncaster Journal & Yorkshire Advertiser*, 25th July 1795
36 Hatfield, Vol. I, p. 420
37 C. Jackson, *Doncaster Charities*, p. 51 for pedigree

Chapter IV Thomas III

1 G. Beard, *Georgian Craftsmen & Their Work*, p. 59 f.n.
2 Hatfield, Vol. II, p. 59
3 Hatfield, Vol. II, p. 34
4 Calendar Doncaster Records, Vol. IV, p. 293
5 Hatfield, Vol. II, p. 74
6 Calendar Doncaster Records, Vol. IV, p. 228 & f.n.
7 Hatfield, Vol. II, p. 59
8 Hatfield, Vol. II, p. 32
9 Calendar Doncaster Records, Vol. IV, p. 148
10 Calendar Doncaster Records, Vol. IV, p. 148
11 Calendar Doncaster Records, Vol. IV, p. 149
12 Hatfield, Vol. II, p. 66
13 Hatfield, Vol. I, p. 31
14 J. Tomlinson, *Doncaster*, pp. 257 & 295
15 These and other references in this chapter are from the Day Book
16 W. Hurst's annotated copy of Edward Miller's *History and Antiquities of Doncaster*, between pp. 192-3
17 *Doncaster, Nottingham and Lincoln Gazette*, 29th May 1846
18 Letter from Muriel Glass to W. K. Sessions, 10th August 1972
19 Hatfield, Vol. I, p. 427
20 Wm. Sheardown, *Moat and Mound. A Defence of the Town of Doncaster*, 1869, p. 32
21 Death Certificate
22 Hatfield, Vol. I, p. 31 f.n.

Chapter V Thomas and Charles Anelay

1 *Doncaster, Nottingham and Lincoln Gazette*, 27th November 1846
2 Hatfield, Vol. II, p. 414
3 *Doncaster, Nottingham and Lincoln Gazette*, 11th December 1896
4 J. E. Jackson, *History of the Ruined Church of St. Mary Magdalene*, p. 9
5 *Op. cit.*, pp. 9 & 10
6 *Op. cit.*, p. 13
7 *Doncaster, Nottingham & Lincoln Gazette*, 4th June 1847
8 W. Sheardown Pamphlet, Vol. I, p. 436, *Doncaster, Nottingham and Lincoln Gazette*
9 *Doncaster, Nottingham & Lincoln Gazette*, 5th October 1860
10 *Doncaster, Nottingham & Lincoln Gazette*, 5th July 1861
11 *Doncaster, Nottingham & Lincoln Gazette*, 18th October 1861
12 J. Tomlinson, *Doncaster*, p. 302. G. Swann, Doncaster Royal Infirmary 1792-1972, pp. 42, 43.
13 Wm. Sheardown, Vol. II, p. 546, *Doncaster, Nottingham and Lincoln Gazette*, 1st September 1865, 19th January, 1866
14 Wm. Sheardown, Pamphlets, Vol. II, p. 643
15 George G. Pace
16 See note 13
17 *Doncaster, Nottingham and Lincoln Gazette*, 17th June 1870
18 *Doncaster, Nottingham and Lincoln Gazette*, 4th April 1856

19 *Doncaster, Nottingham and Lincoln Gazette*, 29th July 1870
20 *Doncaster, Nottingham and Lincoln Gazette*, 16th February 1849
21 *Doncaster, Nottingham and Lincoln Gazette*, 19th October 1855
22 Hatfield, Vol. I, pp. 193/4
23 D. Ware, *Short Dictionary of British Architects*, pp. 152/3
24 A. Birrell, *Life of Sir Frank Lockwood*
25 C. Wilson & P. Pitman, *Encyclopaedia of Murderers*, pp. 507/8
26 J. S. Curl, *Victorian Celebration of Death*, p. 181
27 East Laithgate; Scott Lane
28 Doncaster East Ward, pp. 39, 169
29 Doncaster East Ward, pp. 91, 124
30 J. Tomlinson, *The Lord of Hatfield Chase and Part Adjacent*, p. 220
Doncaster, Nottingham and Lincoln Gazette, 20th August 1875
31 *Doncaster Chronicle*, 21st December 1917
32 *Doncaster, Nottingham and Lincoln Gazette*, 23rd May 1862
33 Hatfield, Vol. II, p. 293
34 *Doncaster Gazette*
35 *Doncaster Chronicle*, 21st December 1917

Chapter VI William I and Thomas Anelay VI

1 *Doncaster, Nottingham and Lincoln Gazette*, 19th February 1875
2 *Doncaster, Nottingham and Lincoln Gazette*, 26th February 1875
3 P. Ferriday, *Lord Grimthorpe*, p. 194
4 P. Ferriday, *Lord Grimthorpe*, p. 132
5 *Doncaster Gazette*, 24th December 1886
6 *Doncaster, Nottingham and Lincoln Gazette*, 21st August 1891
7 J. G. Lockhart, *Viscount Halifax*, Vol. I, pp. 41/2, Vol. II, pp. 93/4
8 *Doncaster, Nottingham and Lincoln Gazette*, 7th July 1893
9 *Doncaster Chronicle*, 19th October 1894
10 *Doncaster Gazette*, 18th May 1906
11 *Doncaster Gazette*, 10th December 1909
12 W. De L'Hôpital, *Westminster Cathedral and its Architect*, Vol. II, p. 337
13 W. De L'Hôpital, *Westminster Cathedral and its Architect*, Vol. II, p. 492
14 *Doncaster Chronicle*, 17th May 1918
15 *Doncaster Chronicle*, 17th May 1918
16 The Builders Journal and Architectural Record, 25th July 1910, p. 455
Also Charles Carus, *Walter Henry Brierley*, 1862-1926, *York Architect*—Diploma in Conservation Studies 1973 (unpublished)
17 Sir N. Pevsner, *Buildings of England, West Riding*, p. 507
18 Anelay MSS.
19 Henry Stapleton, *The York D.A.C. The first five hundred meetings* (unpublished MSS.)
20 Anelay MSS.
21 For a list of the Firm's contracts see p. 62
22 Carus. op. cit., p. 23

Chapter VII William Anelay Limited

1 Yorkshire Archaeological & Topographical Journal 1889, Vol. X, p. 276
2 Anelay MSS.
3 *Doncaster, Nottingham and Lincoln Gazette*, 14th May 1847
4 *Doncaster, Nottingham and Lincoln Gazette*, 29th July 1870
5 Reproduced in *Doncaster Gazette*, 16th May 1960
6 *Doncaster, Nottingham and Lincoln Gazette*, 8th July 1887

15 (above) *St. Wilfrid's, Cantley, Doncaster, restored, a new aisle added and the interior refurnished by William I, 1894. Architect, Sir Ninian Comper. The church was damaged by fire in 1906 and the same architect and builder restored it to its present condition.*

16 (left) *York City and County Bank (now the Midland Bank), High Street, Doncaster, built by William I, 1898, at a cost of approximately £10,000. Architects, James Demaine and Walter Brierley.*

17 *The Principal's House for the School for the Blind (now part of the University of York) at King's Manor, York, built by Thomas VI, 1900. Architect, Walter Brierley.*

PART TWO: THE ARCHITECTURAL PERSPECTIVE FROM THE 1740s TO MODERN TIMES

by George G. Pace, C.V.O., M.A., F.S.A., F.R.I.B.A.

THE ANELAY working span has coincided with a very interesting, even extraordinary, period in architectural history: a period the like of which has never before existed. It may be of interest to make a brief architectural appraisal. In doing so it should be recalled that during this period, not only did the professional architect as we know him today finally emerge, but so did the building contractor. Throughout the Anelay *floruit*, c. 1740-today, the contractor has evolved to become a more and more important member of the building team and, especially between 1850-1938 he was, in a very special sense, the extension of the architect—his hands and in many respects his head too. Whilst the Anelays started working at the tail end of the last phase of living organic architecture in England, they have for the most of their long and distinguished record been concerned with translating into physical reality the dream world of Evocative Architecture, in which lived the building owners and the architects they have so faithfully served.

It may be assumed that unadulterated organic culture (that is, a total culture, unselfconscious, grass-rooted, and functioning within a Closed Tradition) came to an end with the conscious taking of thought (a largely intellectual approach), which brought into existence First Palladianism. But with Second Palladianism, Lord Burlington, together with architects Colen Campbell and William Kent, started something far more radical in principle than anything which in his time and with his background Burlington could have conceived. This growth and the ultimate harvest could not have happened unless all the other essential factors had been present and were to grow stronger and stronger, until not only would organic culture be overcome but could no longer exist. On the surface there does not appear to be much basic difference between the Classical of Inigo Jones and that of William Kent, but by the 1750s we are able to see that the seed of decay had started to burgeon. 'In England it was the world of the spirit, the "interior castle", that was left to decay while the discoveries of scientific materialism were pursued . . . at the price of the empire of the Soul'.*

Evocative Architecture has long roots in England. Such architecture seems to appeal to certain traits in the national attitude to the Arts in general. Vanburgh in a memorandum of 1709 on the arguments for preserving the ancient manor house of Woodstock wrote 'they (ancient buildings) move more lively and pleasing reflections on the persons who have inhabited them'. Earlier in 1642 there is the rather self-consciously argued preference for having Gothic windows in the new Library of St. John's College, Cambridge. Earlier still, in the full flood of High Gothic, the naves of Beverley Minster and Westminster Abbey, and part of the south clerestory of Bridlington Priory, were consciously designed in their main lines in the manner current a century earlier.

---

* Victor Furse, *The Architecture of Sir Christopher Wren*, 1956, p. 159; and Kathleen Raine, *John Donne and the Baroque Doubt* in 'Horizon', xi 66, p. 373

By the mid-18th century some knowledge of the cultural work and artefacts of the past was an essential part of the equipment of those forming Polite Society. The Grand Tour reinforced this and collecting, amateur archaeology, amateur architecture, and especially the desire to recreate a pale, or not so pale, replica of something seen and admired in Europe, became widespread. It was all still rather in the spirit of Hadrian at Tivoli. There continued more and more investigation of the past, more illustrated tomes were published and more and more Architecture of Fashion resulted. The Adam brothers for all their enterprise at Spalato and all their skill, elegance and panache, gave a strong push toward Fashion. Not all was make-belief: serious architecture was conceived in Third Palladianism with the very special and admirable twist given to it by Sir John Soane. But it was permissible for those about to build to choose to do so, if they so wished, in the Indian, the Chinese, the Egyptian, the Gothique. As these exercises were still largely regarded as background make-belief, they are delightful to us for this quality and by the utter lack of concern about mere arid archaeological correctness. But this state of innocence did not last and earnest souls thought to improve by more exact knowledge of the minutiae of detail. A spate of careful measured drawings of the antique Greek, the writings of Johann Joachim Winckelmann and the example of the buildings of Karl Friedrich Schinkel resulted in many a scholarly exercise in the Neo-Grec of the early 19th century. England had its practitioners in this mode: James Stuart, Charles R. Cockerell and William Wilkins. The equally careful delineations of medieval architecture in this country and Normandy brought to an end the charming 'Strawberry Hill' Gothique efforts and ushered in some very dull and painstaking archaeological Gothic. John Nash was not too pleased, 'I hate this Gothic style, one window costs more trouble in designing than two houses ought to do'.

From this point the history of architecture cannot be considered apart from the personalized aesthetic convictions of individual architects, in their writings and projects and, especially in their actual works; the efforts of the lay propagandists; the opening of eyes brought about by the works of certain painters and sculptors, and the effect of engineering works seen through those newly opened eyes.

Time and conditions were now ripe for the introduction of Morals into architecture. This was done with immense energy and a spate of words by such dissimilar protagonists as Augustus Welby Northmore Pugin, John Ruskin and Viollet-le-Duc. It was in new church buildings that much of the polemics and pyrotechnics occured. The Oxford Movement had a tremendous effect on Liturgy and Church life and when coupled with the Cambridge Ecclesiological Movement, an immense impact was made on church building, furnishing and restoration which, in turn, had a marked influence on all departments of architectural thought and action. In secular building before long those who swore by Gothic were locked in battle with those who felt that the Classic (as understood by mid-Victorians) was the only possible vehicle for the 19th century. It was all very real to those waging the Battle of the Styles, but today we see their buildings, whether 'Gothic' or 'Classic', as being quite definitely, utterly and unmistakably, Victorian.

In all this stylistic turmoil two types of architect were designing buildings having rather special qualities, qualities which could not have occurred in the works of any previous age. These are the buildings by the 'Top Architects' and by the 'Rogue Architects'.† Not only did they have a special place in High Victorian Architecture, but they had a part to play in the emergence of Modern Architecture. The 'Top Architects' include: William Butterfield—All Saints, Margaret

† *Rogue Architects of the Victorian Era* was the title of a paper read by H. S. Goodhart-Rendel at the Royal Institute of British Architects, March 1949 and published in the Institute's Journal in April that year.

Street, London; Keble College, Oxford; Baldersby Church, North Riding, Yorkshire; George Edmund Street—Fimber, East Riding, Yorkshire; St. James the Less, Westminster; SS. Philip and James, Oxford: William Burges—St. Finbar Cathedral, Cork; Skelton Church, West Riding, Yorkshire; Castell Coch, Glamorgan; Studley Royal Church, West Riding, Yorkshire; James Brooks—St. Columba's and St. Chad's, Haggerston, London: John Loughborough Pearson—Dalton Holme Church, East Riding, Yorkshire; Appleton-le-Moor Church, North Riding, Yorkshire; St. Michael & All Angels, Croydon; Alfred Waterhouse—Manchester Town Hall.

The 'Rogue Architects' include: Edward Buckton Lamb—Aldwark Church, North Riding, Yorkshire; Christ Church, Hartlepool; St. Mary Magdalene, Addiscombe, Croydon: Bassett Keeling—St. George's, Campden Hill, London; St. Paul's, Upper Penge, London: Samuel Sanders Teulon—St. Paul's, Hampstead, London; St. Paul's, Bermondsey, London.

Towards the end of the 19th century and up to 1914, Gothic in the hands of certain outstanding architects was being handled in an architectural manner and with a degree of refinement probably greater than that normally achieved, or desired, by the originals in Medieval times. George Frederick Bodley—Hoar Cross Church, Staffordshire; St. John's, Cowley, Oxford: John Francis Bentley—The Holy Rood, Watford; St. Luke's, Chiddingstone Causeway, Kent; Temple Lushington Moore—Sledmere Church, East Riding, Yorkshire; St. Wilfrid's, Harrogate: Sir Ninian Comper—Cantley Church, Doncaster; St. Mary's, Wellingborough, Northamptonshire; St. Mary Magdalene, Baker Street, London: Sir Charles Nicholson—Epsom Parish Church; St. Laurence, Eastcote, Middlesex: Sir Michael Tapper—Malvern Link, Worcestershire; The Annunciation, Quebec Street, London.

Some architects had toyed with what was called the Queen Anne Style; notably in the mid-19th century William Eden Nesfield, and in the latter part of the century, Norman Shaw. The last decade of the 19th and the first 14 years of the 20th century saw a decline in revived Gothic, and the emergence of a version of the Classical, which when handled by John Belcher, Norman Shaw and Sir Reginald Blomfield was competent and had an opulence we now associate with the Edwardians. There was also a burst of Baroque by Edwin Rickards at Cardiff Town Hall; The Central Hall, Westminster, and Hull School of Art; and by Ralph Knott in County Hall, London. All this was backed by a host of architects producing well-understood and disciplined buildings in a revival of the lesser Georgian vernacular. The great luminary was Sir Edwin Lutyens in the small country houses of his first phase, the large mansions of his middle phase and the highly personal and witty classical of his final development. The inter-war years may quickly be summed up in the expressive descriptions coined by Osbert Lancaster—Banker's Georgian; Stockbroker's Tudor; Pseudish and Modernistic: all representing the nadir of 150 years of revivals, pretence and personalized dream worlds.

But all had not been lost. We have noted the long decline of organic culture. We have seen the special part played by 'Top Architects' and 'Rogue Architects' in High Victorian Architecture. It is now necessary to observe very briefly the basic steps which led to the emergence of what is now usually described as 'Modern Architecture'.

The Industrial Revolution finally banished organic culture. The new climate encouraged, indeed demanded, the emergence of Individualism. The Closed Tradition vanished. Architecture ceased to be the Mistress Art. The Sister Arts fragmented and all went their unfettered ways, exposed to every whim and fancy of rampant individualism and passing fashions. Cultural chaos, and all that this means to architecture, was now manifest.

But this period of chaos was quite different from anything previous. Not only was there the entirely new phenomenon of technological materialism on a vast scale, but it was no longer possible to absorb and digest it. De-humanization in all departments of life and the abjuring of God, and even of humanism, have produced conditions which have no parallel in history. 'We exist scratchily, competitively, in closely packed agglomerations of humanity. We lack in our ordinary lives that close contact with nature which keeps man sane and humble. We create for ourselves in our urban civilizations an artificial society in which all the criteria are human. There is no appeal to nature; which tells us of fate, and of necessity, how they may be accepted. In our urban societies we come to believe that the whole of our urban society will be easy and good if only man were more honest, more competent, worked harder and were prepared to be uncompromising with evil. We imagine possibilities of excellence which are probably not there. We are in danger of disappointment, unhappiness and despair.'‡ Again—'while machines have multiplied, men have decayed. For just as machines, by carrying too far the principles of division of labour, degraded the craftsman into a machine minder, so, just assuredly and far more subtle, the process of specialization has by perpetual reduction destroyed that catholicity of intellect without which civilization cannot survive'.§

The world had become smaller owing to ease of travel. Information and knowledge both of the past and current theories were broadcast in an overwhelming torrent by printed and *illustrated* books, magazines, and pamphlets. Immense strides were made in recording, and disseminating knowledge *about* the architecture of previous ages. The simple unsophisticated Closed Tradition vanished. Everybody knew too much about what the past had done in architecture.

There could not be a definite architectural vision which carried building far beyond the natural and yet remained the natural, the only manner of building. No longer could the manner of building alter as the years pass, by slow inevitable changes from 'within'. Changes now followed thick and fast, but they were from without, the mere 'titivations' of passing fashions.

The whole man could no longer exist. Augustus Welby Pugin grasped that all was not well with Revived Gothic, and in his propagandist books—but hardly in his buildings—he pointed a way to salvation. Several other Victorian architects in letter, and some obscure ones like Thomas Harris even in deed, did likewise, but with little impact, for the devastating logic of Viollet le Duc was denied to the English theorists: and at this time they would have been incapable of, and certainly have been frightened of, following such theories to where they inevitably led. Engineer-architects, such as Thomas Telford, Isambard Kingdom Brunel, and Robert Stephenson, and those far removed from the practice of architecture, Joseph Paxton and the like, showed the possibilities in their actual works, unbacked by any pretentious aesthetic theories. But the architects and art critics would have none of it. The painters became alive to the possibilities, which eyes newly opened to fresh facets of the eternal core of aesthetics could comprehend. The debt modern architecture owes to Manet and Cezanne, to Matisse and Braque, and later to Picasso and Modigliani is immense. Le Corbusier was once deeply involved as a painter and sculptor, and this painter's and sculptor's awareness is ever-present in his buildings. Generally, sculptors were slower than the painters in exploiting the new awareness. But architects, if they are to experiment, need to have opportunities on a very different scale from those needed by painters and sculptors. They have to have a client who is prepared to bear the immense cost of the architect's experiments. This pre-supposes in the client an enlightened interest in new creation in architecture.

‡ Laurence Brander, *George Orwell*
§ L. T. C. Rolt, *I. K. Brunel*

18, 19 *'Bishopsbarns'. Walter Brierley's house for himself, St. George's Place, York, showing the road facade and the staircase hall, built by Thomas VI, 1905. Brierley was so pleased with the work that he sent Thomas VI a silver inkstand and a cheque for £100.*

20 (left) *The workshop and factory in Bishophill, York, built for T. Cooke & Sons Ltd. by Thomas VI, 1916. The firm of Cooke's which is now Cooke, Troughton & Simms Ltd., made vital optical instruments during the 1914-18, 1939-45 wars. Architects, Brierley and Rutherford.*

21 (below) *'Goddards', Tadcaster Road, York, built for Noel Terry by Thomas VI, 1926. Architect, Walter Brierley, who died before it was completed, and James Rutherford.*

In any event, it is much easier for so-called informed opinion on painting and sculpture to be nourished by the spoken or written word, and when it has been so nourished, for many reasonably intelligent people to be able to come quietly to an understanding and appreciation of continuing developments in these arts. New creation in architecture is altogether more difficult to explain even to the intelligent. To very few are given the gifts needed before new creations in architecture may be assimilated, understood, or appreciated; especially is this so at a time when the Historical Conscience is so highly developed. The scales have been heavily weighted against the general appreciation of the possibilities in new architectural creation, and architects have had a long battle for the opportunity to experiment and demonstrate their belief in modern architecture.

But all architects could not be smothered all the time. The works of the early engineer-architects and the painters worked as leaven. New conceptions of the interplay of solid and void, new capital relationships, new awareness of shape, line and silhouette, of texture and colour began to emerge and to be appreciated. In this ever-present witness, railways, ships, plants and machinery of all kinds played a great part. Alongside the scholarly and the eclectic architecture of the 19th century, there existed a despised but vigorous industrial vernacular. Until recently the importance of this industrial vernacular has been rather overlooked.

William Morris, Charles Annesley Voysey and William Richard Lethaby performed vital pioneer work in creating the conditions and climate essential if there was to be general emergence of Modern Architecture in this country. It has been suggested that a more obvious immediate forerunner of Modern Architecture is to be found in Art Nouveau, but this was largely a passing phase, useful though it was in an age of revolution. Here works by Charles Mackintosh, George Walton, Edgar Wood, Arthur Mackmurdo and Harry Wilson, could not stem the return to stylistic revivals which, until the 1930's, were to pass as architecture, in this country; revivals so well-mannered and in such good taste that the redemptive feature of all 19th century revivals—that of being 19th century first and a revival second—was absent.

In spite of the significance of the Glasgow School of Art and the works of Walton, Wood, Mackmurdo, and Wilson, the leadership of Modern Architecture passed to the Continent, and there it has remained. We have not the space to follow the fortunes of Art Nouveau; of the Arts and Crafts Movement; of the impact of the European Architects, Henri van de Velde, Otto Wagner, H. P. Berlage, Peter Behrens; of the Deutscher Werkbund; or of the U.S.A. Architects, Louis Sullivan, H. H. Richardson and Frank Lloyd Wright. In Gropius we may see the catalysis of the New Style. He largely determined the character of Modern Architecture. Sir Nicolaus Pevsner has likened his achievements to the pioneering break-through from Romanesque to fully-fledged Gothic which the Master of the eastern chapels and ambulatory of St. Denis achieved in one supreme white-hot creative act; the root from which all Gothic development of the next 80 years sprang. Against the achievement of Gropius, Gaudi and Sant' Elia appear as freaks and Art Nouveau the invention of raving individualists, quite properly in revolution against the smug make-belief of the 19th century, but without any real understanding of the basis from which the new architecture must spring and the factors it must satisfy. Gropius' 'supreme white-hot creative act', the Model Factory in the Werkbund Exhibition, Cologne (1914), gave aesthetic form to the fundamentals from which Modern Architecture must spring and satisfy—faith in science and technology, faith in social science and rational planning, faith in machines, belief in architecture being a service for all men and that the warm personal relationship between architect and client, as in past organic but craft cultures, has gone for good and that the new conditions require to be cold, so as to command impersonal technical processes and anonymous clients; belief that there is no place in the present world (and thus in architecture) for mystery and other-worldly speculation and, above all, no place for Art for Art's sake.

It will be noticed that after playing an immensely important part in the earliest stages and a potent part in the later part of the 19th century and the early years of the 20th century, this country returned to and occupied itself with the make-beliefs so aptly named by Osbert Lancaster; the works by Wells Coates, Maxwell Fry, Connel, Ward and Lucas, and Tecton, were isolated but brave attempts to use here the developments which had been possible on the Continent. But this country had missed the Continent's many years of discipline and the stripping down to the basic Modern Architecture. It resembled the urgent pianist trying to play Chopin without going through the tedium of the exercise and practice. After World War II, economics and the presence of the second generation of architects from the Schools made sure that the stylistic *semblance* of Modern Architecture would be the normal approach in this country. But as the infantile fashions and clichés which follow one another with bewildering speed are seen, then only sadness can be left that the absence of the tough period of training and discipline has been fatal—the semblance but not the spirit is present.

The Anelay working span has been against this architectural background. They have been serving architects who have held many of the varying architectural philosophies of the period, and most facets of these philosophies and fashions have occurred in the works with which they have been involved.

Cabinet-Maker

PART THREE: THE EARLY ANELAY ACCOUNT BOOKS

A Study by James E. Day

THE SURVIVING Account books of the firm of Anelay are three in number—two small ones which are really memoranda or Day Books, and a larger leather bound folio Account Book.

The First Day Book

The first of these commences on 28th April, 1772 and continues until 30th December, 1774, a period of nearly three years. The pages are not numbered, a fact which renders ready reference to, and analysis of the contents, somewhat difficult. At the beginning of this volume is an account of how Messrs. John Thompson and Henry Walker built a raised causeway in Marshgate for the Corporation of Doncaster. This causeway, or as the bill expresses it, 'raising of Marshgate', was constructed in the year 1753, between the Friars Bridge, which spanned the River Cheswold, and St. Mary's or Mill Bridge over the River Don, and elevated the roadway some 10 feet above its former level. The significance of the bill's inclusion, is that John Thompson was the 'uncle' of the Thomas Anelay (Thomas II, 1739-98) who succeeded to the business on Thompson's death (1771), there being frequent references to 'Ant Thompson' (John Thompson's wife) in the book. The works described in this volume are chiefly of a domestic nature, with two possible exceptions which will be examined later.

There were certain stirrings of civic pride at this date, accelerated no doubt by the building of Doncaster's Mansion House, but possibly the chief spur to the re-modelling of the town's main streets was the advent of the turnpike roads, when the consequent increase in traffic made the demolition of projecting houses necessary. It is apparent that at this period much of the work being undertaken by Anelay, either of construction or repair, was at farms in the villages of Rossington and Long Sandall, where the Doncaster Corporation owned estates. In May, 1772, Mr. Beale's house at Sandall was undergoing 'Plastering and Dashin' and in the same month, the master, with a bricklayer and a lad, was building a wall for the Corporation at Rossington Parks. On 2nd June in the same year, minor repairs were being carried out on their pride and joy, 'Mention house mending' (the Mansion House); the day previously the builder had been working 'at Mr. Holmes Sprotboura Boat'. This was an ancient inn which stood on the north bank of the canal at Sprotbrough; its successor later called 'The Copley Arms', is now the farmhouse of Boat Farm.

On 5th June an entry runs 'To 1. Chalder Lime to Presptearon Chapl', while a later entry speaks of 'Mr. Scoot at Meetin hous'. Mr. Scoot may be identified with the Reverend Andrew Scott, who was then minister of the first dissenting chapel in Doncaster, which was built in 1740 on the west side of Hallgate. The congregation experienced doctrinal changes common to Presbyterians of this earlier period, finally adopting Unitarianism. There will be occasion to refer to the Revd. Andrew Scott and the chapel later. (See page 40.)

In August, Anelay's men were engaged for a number of days on a wall at the Mill bridge, and in 'filling hole at Mill Wassh'. The Corporation accounts contain an entry which must refer to this commission, 'Oct. 29 Part of the Expc of a Foundation and Wall in the River Opposite the Mills as by Accot, T. Ainley's men being charged in his Sundry Bill—£6.4.1½'[1] Mr. Pigot's farmhouse at Rossington was at this time rapidly nearing completion, '21 to 26. At Pigot's Rossington at Gables and Chimney 2 days'. In the Corporation Cash Book the provisions for the rearing supper are detailed 'Sept 5 pd. for a piece of Beef for Wm. Pigotts Rearing at Rossington—9/6. To Ale for 26 persons @ 8d. apiece 17/4. and Allowed for Bread, Roots, etc. 3/8—£1-1-0'.[2]

One of the two relatively major works undertaken by Thomas II during the period covered by the Day Book, was commenced on 5th October, 1772. The job was the arching over of an open sewer. Here it may be of assistance if a look is taken at medieval Doncaster. From the late 12th century the small town had been encircled by a defensive ditch, the course of which is roughly indicated by existing, or lately existing streets; primarily the ditch branched from the now culverted River Cheswold at a point near the present Northern Bus Station, followed the line of Factory Lane (now almost totally demolished), crossed St. Sepulchre Gate continuing along Printing Office Street to its junction with Cleveland Street. Here the ditch turned eastward, crossing High Street, from thence it passed down Silver Street and Market Road where it joined the River Don, the course of which in comparatively recent times (c. 1934) has been diverted, the old river bed being filled in. In the course of time the greater part of the ditch (known as the Bardyke) had also been filled in, but the portion from Sunny Bar to the river remained open, and had been degraded to serve as a public sewer. This state of affairs had existed since the late 17th century. An entry in *The Courtiers* (28th May, 1721) refers to the paving of a channel at the back of the town hall, 'to carry the water thence down into the Barr Dyke sewer'.[3] When the Corporation determined that the sewer should be enclosed we do not know; the printed Calendars make no reference to any resolution on the subject. Improvements in the Market Place area were contemplated in the mid-18th century and it is possible that the open water course was an obstacle to the project. It may be that the sewer was becoming increasingly offensive even to the 18th century nose.

The first reference dated 5th October, 1772, is rather a cryptic one. It reads, 'for Corporation at saw in healeys Croft'. Translated into the modern idiom it reads 'for Corporation at Sough in Healey's Croft'; the word sough is an ancient equivalent of drain, and the Healey referred to was Alderman George Healey (died 1773) who was a tanner and whose yard was near the water-side in Fishergate. There are two entries in the Corporation Cash Book which refer to the work; on 8th October 'Wm. Morton, paid for Cutting plank for the Soughs in Fishergate as p. 12/4', and on the 22nd of the same month a further payment to Morton is entered 'for cutting 60 yards new Oak plank for the Soughs in Fishergate by order of Mr. Rickard—7/6'.[4] It is suggested that this timber may have been intended for the arch centerings, to be employed in arching over the ditch.

Thomas II employed a large staff on the work at various periods extending over some four months, the localities in which the work was proceeding being recorded in his Day Book. What appears to be the last entry dated 1st and 2nd February is headed 'for Corp at Saw in Parsnidge'. This referes to the area long known (and used by the tithe farmer) as the Parsonage Yard. It was later acquired by the Corporation for the building of a cattle-market. In the Cash Book of 5th May, 1773, there is a payment to '5 Labourers 10 Days filling Ribbish to back the

Sough on the Back of the parsonage 15/-'[5], presumably this refers to the subsequent infilling of the arched sewer.

Wm. Sheardown refers to the enclosing of the Bardyke in one of his historical pamphlets. He writes 'Mr. Anelay, a respectable and intelligent mason, who died on 19th July, 1848, stated he well recollected the ditch from Sunny Bar to the river at Docken Hill being open, that it was partly within the walls of the parsonage yard and that his father arched it in.'[6] There is one other reference to the sough in the Cash Book, a rather late one, but it is possible that payment was delayed. 'Nov. 20 1773. Reuben Marsden for a Scope that got Broke at Mr. Healey's Croft'.

Despite his preoccupation with this large commission, the builder did not neglect his humbler customers. On 14th October we find him 'At Mrs. Whits, Cleaning flew of oven' for which he charged 6d. In the following month of November (1772) two men, Varley and Wild, had to be diverted from work on the sough to carry out repairs at the Workhouse. This stood on the south side of St. Sepulchre Gate, near the bottom of Printing Office Street. Originally it had been the home of Alderman Pell, but had been acquired by Miss Jane Ellerker, who in her turn had sold it to the Corporation in 1730 for use as a House of Maintenance. The entry runs 'Nbr 7 at Workhouse Mending Chimney'; the 'Mending' might more adequately have been described as rebuilding, for the men were engaged at the work for six days, using no less than 650 bricks and 21 barrows of mortar.

Three jobs recorded in December are of interest, the first of them for 'Mr. William Crowder, Senior at Stove &c.' Mr. Crowder was a nurseryman and the 'Stove' mentioned was in reality a hot-house (they were often described as 'Stoves'), the two men employed were paid 8d. a day for ale; no doubt, at this time of the year, Mr. Crowder would be concerned for his plants, and the ale would be intended as an inducement to the men to complete the work as speedily as possible. The second job was like the first, but this time for 'Wm. Crowder, junior', who also gave the men liquid encouragement. The third task referred to was undertaken for 'Mr. Broadhead, Esq.', the work consisting first of putting in a door, and several days later of the pointing of the pediment and re-pebble-dashing. Mr. Broadhead was of sufficient local importance for Thomas II to include himself in the staff engaged. Mr. Broadhead's mansion was situated near the Mansion House and was later bought by Wm. Sheardown, who converted the stable block into a printing-office for the *Doncaster Gazette* of which he was the then proprietor. The house was subsequently demolished, Priory Place being erected on part of the site; half of the original stable block, however, still remains and is now occupied as a shop by Messrs. Dibb and Waddington, drapers. The entry in the Day Book is, we believe, the only reference to the fact that the Broadhead house had a pediment.

In mid-December work was still proceeding on the sough, and at the same period the builder had a gang at work for the Corporation at William Foster's farm at Sandall, while Anelay and a bricklayer were working three days 'At Joseway Broadheads at fatts & Coppers'. Joshua Broadhead was a dyer, with premises in Marshgate.

At the end of this year and the beginning of 1773, Thomas II experienced some financial difficulty and had almost weekly recourse, either to 'Ant Thompson' or to Thomas Rimington (a grocer and near neighbour), for a loan. It must be remembered that he had but recently succeeded to the business and could have had few financial resources; moreover, he had to pay the weekly wages to his men, even if he obtained his materials on credit. Unfortunately the Day Book gives no weekly rates of payment either for bricklayers or labourers, so his out-goings cannot be ascertained in this direction. The chief cause of his troubles was undoubtedly dilatory payments on account by the Corporation, his principal employer; whether this was due to laxity on the part of the steward, Charles Bradford, or a shortage of money in the Corporation coffers, is not easy to determine. From 17th October, 1772, when Anelay received £10.10.0, to 16th February, 1773, when he was paid £30, there are no payments recorded in the Corporation ledger. The same absence is also apparent in the accounts of other tradesmen working for the Corporation at this time, notably William Rickard, who had not been paid anything from 21st May, 1772, until 16th February, 1773, although, as he had been successfully established in the carpentering trade for many years, he was the better able to meet his commitments. It may be of interest to remark that in the June of the year 1772 the Corporation had been able to find sufficient money to purchase a robe and gold chain for the Mayor, and to pay for the gilding of the great mace, at a total cost of £105.3.6. A further payment of £74.2.1 on 18th February[7] released Thomas II from his immediate financial difficulties, as his total borrowings had only amounted to £26.5.0; indeed, in the ensuing month he was able to lend his aunt eight guineas, in addition to the repayment of her loan.

On 1st February he records that two of his men were 'at Mr. Bowker's Setting Kettle' for which job they used 30 bricks and two barrows of mortar; the kettle referred to was a hatter's kettle, for Mr. Bowker was a Doncaster hatter, 'February 12, 1773 Jonath Bowker for 3 Hatts for the Bellman, Bedals, & Pinder as p. — £1.5.6.'[8] The £30 previously mentioned was paid on account for work at Rossington; work was still proceeding at Pigot's Farm, where plaster floors were being laid, a job which eventually entailed the employment of a large work force.

At this point it may be advisable to furnish a short explanation of the methods used to prepare plaster at this period. It was necessary that the raw gypsum should be calcined before it could be used, the process being carried out in a roughly constructed kiln erected on the building site; if any quantity of plaster was required, this roasting was often a lengthy operation which necessitated men working throughout the night tending the working of the kiln. After the gypsum had been burnt, the calcined nodules had to be reduced into powder form. This was accomplished by the use of the flail, a primitive threshing implement, which consisted of a long cylindrical wooden handle with a swivel fixed to its top; to this swivel was attached a pole of similar length to the handle, the implement being used with a beating motion. 'Nov. 2, 1813. By Cash pd for a quantity of Flail Swivels & Bands for Thrashing Plaster—5/-'.[9]

Work at the Pigot's Farm floors lasted from 15th to 20th February, two kilns being set up. 'Setting up at Nights 2 men. 4/-'; the keying material here used consisted of a layer of reeds, 'Reeding floors—3/6'. These floors had a smooth finish 'Smoothe floor 1/-'. For the men who 'Set up at Night' candles were provided at a cost of 2d—and as of necessity it was a 'thirsty job' there were copious supplies of ale. Much of the plaster used by the Corporation at this period came from the gypsum pits at Cusworth, Nr. Doncaster. 'April 5th 1788 Wm. Wrightson Esq. for 5 Tons Brown Plaster to Rofs [Rossington] as p. Rt. Storrs—£1.5.0'.[10]

In the following month the builder and his men were employed paving the weir at Sprotbrough, '31st March 1773 for Mr. Holmes paving wash at Sprotbourough'. The weather was not very kind, for Thomas II made a note in the margin of his book 'frost and snow'; there are also references to ale. It should be noted that when the term 'paving' occurs in accounts of this date, it usually denotes paving with cobble stones, and does not refer to the laying of flags. Over 1,400 tons of cobbles were bought to pave the Doncaster streets over a period of 11 years, being transported to the town by water, sometimes brought from Ripon, but often by ship from Scotland to Hull, where they were transhipped into keels, for the remainder of their journey.

Earlier in this study, contemplated improvements in the Market area have been mentioned. One of these was undertaken by Anelay; it was the provision of a wall in the Market Place, or more particularly, in the Corn Market. In this locality there stood a stone column surmounted by a ball, known as the Wheat Cross, which had, at its base, a circle of flags on which the corn sacks were pitched. The original intention of the Corporation had been to replace the Cross by a new one, and stone had been dressed by Edward Jackson for this purpose. An entry in *The Courtiers* dated 1st June, 1770, orders that he shall be paid for this work[11]; the entry refers further to the intended pulling down of houses in the Market Place, and to the building of a wall if thought necessary. It is apparent from the above that the area devoted to the sale of corn was becoming inadequate to meet the needs of an expanding corn trade, and that a wall would provide more accommodation for the pitching of sacks than would the erection of a new cross. The first entry relative to this wall in the Day Book is dated 18th May, 1773. Certain details of the work appear, getting bricks on to the site, cutting arches, the backing of the plinth, etc. As far as can be ascertained the work continued without interruption until the beginning of August, the number of men employed on it varying as the job proceeded. An entry dated 5th May, 1773, runs 'Taken Down Pear &c'. Almost certainly this entry refers to the demolition of the old Wheat Cross, which a builder would naturally describe as a pier; this is the only reference seen hitherto which gives a definite date and year for the disappearance of the Cross, and is of great interest to the local historian.

It is suggested, from the references to arches, that it was a wall which had a succession of blind arches along its length, for the accommodation of corn sacks; Dr. Miller mentions the wall as 'an ornamental wall'[12] which he says 'was erected in the Corn Market at the expence of the Corporation'; it should be stressed that the above remarks as to the nature of the wall are only speculation, but nevertheless they have not been made without reason. The worthy Doctor would appear to be in error in his recording of the building of the wall, which he says was built in 1772, whereas the Day Book entries are all dated 1773, save one which occurs 'Apr. 26 1774 for Corpation pointing Arches in Market Place'. It is probable that the corn wall is intended and that the arches were its most prominent feature. During the wall's progress, from 21st to 26th June, part of the staff were diverted to the 'pulling down house/Markitt place &c' while 'Slack & Lad' were still 'walling & Cutting Arches'. The last entry for the building of the wall is dated 7th August, 1773, Thomas II's Bill 'for Market Place Wall' amounting to £27.16.3.[13]

An entry dated 9th October, 1773, refers to the master builder's horse, 'put Mare to grafs at Mr. Gibson/took out the l. of November'. In the January of 1774 Thomas is taking up the paving in Marshgate for the Corporation, and 24th of the same month he records his hay transactions with brother Charles, and also details of the shoeing of the mare '6 new shoes & 2 removes'. The following May a small force is engaged 'Reparing Vestery parte of Rossington Church' and 'at Church Mending Steple & Leds'.

The Butter Cross, at the Baxtergate entrance to the Market Place, was evidently in need of repair in the month of August, when Anelay and a bricklayer were engaged in 'Scaffillding [Scaffolding] the Cross'. The remainder of the year follows the familiar pattern of small commissions, for which there is not space to examine in any further detail.

The Second Day Book

From the close of the first Day Book on 30th December, 1774, there follows a dark period until the opening of the Large Account Book on 27th February, 1798.

The second Day Book need not involve us in any lengthy examination, for most of the information recorded between its covers has been extended into the large Account Book; this second Day Book covers a period of approximately seven months in the year 1800 ranging from 10th March to 20th October.

There is one item of interest in the closing pages of the Book; a detailed list of Thomas Anelay II's transactions with Dr. Miller (the Doncaster composer-organist and historian) relating to the supply of building sand from the latter's close. The list ranges from 3rd March, 1798, to 9th May, 1800, during which time the builder received 41 loads of sand from Dr. Miller. After the latter date Anelay obtained five loads of sand from a Mr. Gandy; the last load is undated but the date of supply is after 7th November, 1800, when the list appears to have been discontinued.

The Large Account Book

This extends over a period of approximately 44 years, from 1798 to 1842. The entries are not in chronological order; the book-keeper appears to have worked through to the back page, but in many instances blank pages have been left which have been utilized at a later date. The entries vary in interest, some being more informative than others, but the book is unquestionably of great value to the local historian, especially in cases which can be identified in the Doncaster Corporation Accounts.

Also of great interest is the number and status of the builder's clients. They varied, as may be expected, and ranged from local gentry (such as the Cookes of Wheatley, and the Copleys of Sprotbrough), through the professional and tradespeople of the town to quite humble customers. The Corporation was the builder's principal employer, and while comment must be made on some other interesting private accounts, and the people to whom they were addressed, it is the Corporation items which are the most rewarding.

The Book opens on 27th February, 1798 (three days after the death of Thomas II), with a commission from Doncaster Friends (to quote, 'the people who are commonly called Quakers') for making a Meeting House. It is possible that attendance at the Warmsworth Meeting House was becoming more difficult for the older members of the Society, hence the decision to make a Meeting House in Doncaster.

C. W. Hatfield refers to the opening of the Meeting House on 22nd July, 1798, and quotes a Mr. Thomas H. Harrison's opinion that it had been originally a barn.[14] The Account Book (fol. No. 2) in fact refers to it as a granary; the structure of the original building was considerably altered, windows and doors being 'broken out', and ceilings made and plastered, both in the Meeting House and in the keeper's cottage, laths and reeds being specified for this latter purpose, as well as '9540 Lath Nailes at 1s-5d pr Thousand'. The exterior was pebble-dashed (for which 5 Metts of pebbles were bought), and subsequently coloured.

On folios 56 to 59 appears Mr. John Robinson's account which extends from the year 1798 to the year 1803. Robinson was a wharfinger and also the lessee of the ancient water-tolls from the Doncaster Corporation. The account contains several items of interest, among them one for 'Making Wheel pitt'. This was in all probability a pit for the bone mill, used for grinding bones

22 (above) *The dual purpose Church/Hall of James the Deacon, Dringhouses, York, built by William II, 1955. Architect, George G. Pace.*

23 (below) *The Welch Regiment Chapel, Llandaff Cathedral. Furniture by William II, 1956. Architect, George G. Pace.*

24 (above) *The County Assize Courts, York, originally designed by John Carr. In 1958 William II carried out major repairs and alterations, including renewing the columns to the portico. Architects, Kenneth Ward and Cecil G. F. Ward.*

25 (below) *Belton House, Grantham, Lincolnshire, where William II undertook a major restoration in 1962. Architect, Francis F. Johnson of Bridlington.*

into bone meal, then extensively used as a fertilizer. On folio 57 there is a reference to alterations to the Bone-house. Other items relate to the boatyard and to the keelyard, keels being built on the premises; in this connection the following extract (fol. 59) is worthy of remark. 'Oct, 29 1803 Setting Range in Vessel $\frac{1}{2}$ Day Bricklayer and Labouring 3/1$\frac{1}{2}$'. This would be the range and its accompanying brickwork set in the tiny living cabin of a Humber keel.

'Mr. Richard Anelay's Day Accompt' on fol. 58 relates to a public-house in Fishergate known as 'The Crown'. (Mrs. Ainley was the owner in 1806).[15] There are references to 'the Brewhouse'; to 'altering Barr' and to 'Building Gantry'. Another item in 1799 (fol. 133 extended from fol. 58) runs 'Removing Bricks from Factory Wall. Walling up door'. It is probable that the 'Factory' here alluded to, was the weaving mill built by the Revd. Edmund Cartwright, the inventor of the power loom, whose venture in Doncaster had failed, the factory being advertised for sale in the year 1794.

Among the larger works appearing in the Account Book is the alteration of the Mansion House front, when an attic story was substituted for the existing centre pediment; the projected side pavilions, designed by James Paine, were never constructed, and the pediment appeared very oppressive in consequence.

William Lindley, who had been in John Carr's office in York (see page 62, reference 1801), and who had been responsible for the design and construction of the Doncaster Play-house, as well as the Doncaster Dispensary, had contracted with the Corporation to complete the alterations to the Mansion House (to his accepted plans) for the sum of £950.[16] On fols. 187 to 189 is Mr. Lindley's Account which is headed 'Mr. Lindley's Day Accompt of Repairing Mansion House'. It is apparent that the architect had sub-contracted the necessary brickwork, plastering, pebble-dashing, etc., to Thomas III. Probably it should be explained that at this period it was quite an accepted practice for an architect to contract for the erection of a building which he had designed, although it was not an invariable rule that he should do so; i.e. the building of the Dispensary was a direct contract between the Corporation and Thomas II.

The work of alteration extended over a period of some seven months from 23rd February, 1801, to 22nd August, the contract date for completion being 15th August, a stipulation obviously made because of the imminence of the Doncaster Race Week, with its attendant assemblies and balls. Thomas III's account for the work is very incomplete, and enters into no great detail regarding the work performed; the number of men employed in each week is given, but contrary to the builder's usual practice in Day Accounts, their names are omitted; it is possible that the reason for this lack of detail is that the work would be closely supervised by the architect. On folios 188 and 189 is 'Mr. Lindley's Accompt for Materials'. This is of great interest, although sadly incomplete, as many of the prices and cost are not given, but the account supplies the information that the builder was also responsible for extensive interior decoration of the building at this time. As the work progressed, payment on account was regularly made by Mr. Lindley, who being very much of 'the gentleman', always paid in guineas.

It is obvious from the entries in the book, that Thomas III and his workmen were continuously employed in the maintenance and repair of Corporation property. Folio 16 dated 1798 contains several items, such as repairing the slating on the Dispensary, mending the tiling on Corporation houses, whitewashing and colouring stand kitchen and the repairing of town drains. The winter and spring of the year 1799 would seem to have been exceedingly unkind and the builder's men were requisitioned to clear the snow from the roofs of Corporation premises. In the

Day Account for that year on 4th February the entry runs 'To Throwing Snow of the Mansion House—2/-; Do of Playhouse—1/6; Do of Town Hall—1/6.' Similar sums were paid on 5th April, but whether there was a heavy fall of snow on that date or whether the entry was belated is not clear. The practice of clearing the roofs of snow was one of long standing, for the Corporation accounts record the payment of 1/- for 'Two Old Malt Shovles to throw Snow of M. H. Theatre and Stand', this on the 27th November, 1786.[17] Later items in 1799, included work on the Grand Stand, such as tiling, slating, mending plastering, pointing tiles and walls, and whitewashing and colouring, while a new shed was provided for the town jail.

The February of 1800 (fol. 145) sees Anelay engaged on substantial repairs to the Town Hall (then situated in the Market Place); the slating was repaired, and also the pebble-dashing of the exterior, the outside of the building was 'collered' and also whitewashed and coloured. Repairs to the interior were undertaken with some urgency, for 16d was paid for candles. Evidently the accommodation for the 'twelve good men and true' was inadequate, for in April the entries read '7 to 12 Stopping plastering, Breaking Down wall for Jewry' and on fol. 146, 'April 14 to 19 Altering Jury place'. The interior of the hall was also coloured in both green and yellow wash, while 30 yards of cornice were whitewashed. At the same period the front of the playhouse was pointed, for which 20/2d. was charged, with a further 3/- for fine lime; the completion of this work in the Market Place is marked by an entry 'April 12 Mending paving where Scaffolds stood, 1 day Loyde, Pearson—4/4.'

The scene of the builder's labours moved at the end of June to the Waterhouse which stood on the south-east side of the junction of Bennetthorpe and Thorne Road. This had been built in the year 1794 at the then highest point of the town, the water being lifted into the wooden reservoir housed inside it, by pumps driven by an undershot water-wheel situated on the River Cheswold. There is substantial evidence to indicate that the reservoir had always been subject to continuous leakage, for only three years after its construction the Corporation Accounts show that two ship carpenters were employed one day at the reservoir caulking the seams with oakum;[18] this leakage would naturally affect the house structure with the passage of time. The entry marking the commencement of repairs by Thomas III is on fol. 146 and runs '1800 30 June to 5th July. Taking of[f] Dashing, picking joints, making Tarras—'.

It is evident that the constant leakage from the tank had seriously affected the mortar joints of the Waterhouse, so much so that drastic and expensive measures were deemed essential for its repair. The pebble-dashing was stripped from the building to give access to the brick joints; these were then raked out, removing so much of the decayed mortar. This operation no doubt would be carried out both on the inside and outside of the building. The joints were then pointed using Terrass, a species of hydraulic cement much used in the construction of canal locks and similar works, by reason of its water resistant properties. The builder employed a numerous staff on the work for some three weeks. The house then appears to have been left for the same period, then white-washing is recorded (this was probably carried out on the interior). Its exterior was finally coloured in the beginning of September. The Corporation's architect was evidently dissatisfied with the result for the building was again coloured, by his order, a fortnight later.

After the 'tumult and shouting' of the Race Week had subsided, we find the builder's men again at the Waterhouse (fol. 147) 'Oct. 13 to 18 At Waterhouse plastering Suffeats. 2 Days. Thrush 2/8. Labourer 2 do 2/3—9/10. 3 Barrows Lime & hair at 1/10—3/-'. The term 'Suffeats' was initially rather puzzling, but it is suggested that the book-keeper intended it for Soffits. It is conjectured that this may refer to the semi-circular arched recess in the main or South Parade front of the Waterhouse, and that it was plastered in contrast to the pebble-dashed surface of the rest of the building.

One small commission executed towards the end of the year 1800 will be of interest to the social historian. *The Courtiers* contain an entry dated 8th October, 1800, to the effect that several of the principal inhabitants and societies of clubs had agreed to subscribe money bearing interest at 5 per cent in order that a fund might be raised for buying corn to be ground into meal, and that it might be sold by retail without profit, excepting the payment of the interest and working expenses, so that the poor of the borough might be served upon the lowest terms, while the price of corn remained high.[19] The committee of gentlemen-subscribers appointed to manage the business lost little time in carrying the proposals into effect, for on folio 32 of the Anelay Account Book occurs the 'Meal Club Day Accompt' in which the first entry is dated 27th October, 1800. Where the shop was situated is not mentioned, but it emerges from the account that the building needed certain minor repairs, such as slating and tiling and also the renewal of some interior plastering, for which work the builder charged £12.11.0. That these measures were opportune is evidenced by a further resolution of the Corporation '23 December 1800, Ordered that the corporation contribute £500 towards the relief of the poor in Doncaster'.[20]

In the January of 1801, among a number of small repair jobs, Anelay's men were employed in the removal of lead from the Butter Cross, but on whether this was an emergency measure, or the lead was taken down with a view to its ultimate renewal, the accounts are silent. In the month of February minor repairs were made to the Town Jail which included the renewal of two ridge tiles at 2d. apiece. From 23rd March to 28th April the work scene moves to Rossington where the Corporation brick kilns were partially re-built; these kilns (which were in operation as early as the year 1772), usually commenced their burning season in mid-May and operated until about September, when some 14 to 16 kilns of bricks were made. In the four years 1786 to 1789, the total output was 1,550,349 bricks.[21] The Corporation steward solemnly rode to Rossington at the end of each burning to check the number of bricks taken out of the kiln. He rode a hired horse at a cost of 1/6;[22] the necessity of this exercise was occasioned by the imposition of a Brick Duty in the year 1784 of 2/6 per thousand.

The Corporation had found occasion to increase the rents of their Rossington tenants who, in their turn, had requested that the open arable fields might be enclosed in order to improve their farms, a proposal to which the Corporation agreed. An obstacle to the carrying out of the scheme lay in the disposition of the rector's glebe, and it was ordered 'that a committee treet with Mr. Stovin, the rector, for making an exchange of such lands so that they may be laid together'.[23] From the same source we learn that the Rossington committee were ordered to 'cause a school house and a school room to be built at Rossington fit for the residence of a schoolmaster', the order bearing the date 30th April, 1801. Nothing is heard of this resolution for several months. In the interim Mr. Anelay's men are again employed in re-colouring the Waterhouse by Mr. Lindley's order (fol. 146); perhaps his solicitude in the matter may be partly explained by the circumstance that his three newly erected houses on South Parade faced the offending building. The houses, colloquially called 'The Pillared Houses', still ornament South Parade. The first intimation that work had begun on the new school house is contained in the 'Corporation Day Accompt for Rossington—1801' (fol. 206); the opening entry is dated 27th to 31st July. This entry is uninformative, but in the ensuing September the entries refer to the digging of cellars and foundations. The Measured Account for the school house (fol. 208) commences on 5th to 10th

October and continues to 4th September, 1802; the monetary details are somewhat obscure, as certain items are included which obviously relate to work at Mr. Jackson's, that is at Holmscarr, or Rossington Grange. The new building had some little architectural pretentions, for in the Day Account we read 'Labourer assisting Masons' (fol. 206); 'Getting Cornice out of Carts'; 'At Rossington School, Building pediments, Assisting Masons, &C &C' (fol. 207). The Rossington School items are in a general Corporation Day Account, which contains others which refer to work in Doncaster, and it has not been thought worthwhile at this stage to abstract the details, especially as the Corporation accounts invariably contain block payments and would be of no confirmatory assistance.

Reference has previously been made to the dissenters' chapel and its Minister, the Reverend Andrew Scott, who had been tutor to the children of Sir George Cooke, Baronet, of Wheatley Hall. This gentleman in 1797 resided on the east side of Hallgate, in a house nearly opposite the chapel. (See page 31.)

On the second fly leaf of the large Account Book there is an 'Accompt for Altering a house in Hallgate Belonging to Sir George Cook, Bart. of Wheatley, now occupied by Mr. Thos. Coulter'; details of the work are scanty, but evidently comprised repairs to the Stable, Coach-house, and pig-coat, and extensive re-tiling; the whole of the account only amounting to £8.11.5, of which Sir George Cooke paid £2.4.0, presumably leaving the balance to be settled by his tenant. This house may be identified as the one previously occupied by the Revd. Andrew Scott (he died in 1799), the site of which has now been absorbed by an extension to the 'White Bear' public house; it is a matter for speculation whether the baronet provided a house in Doncaster for his children's tutor in lieu of quarters at Wheatley, in any case the house was not occupied by Scott's successor, the Revd. Richard Hodgson. During the latter's term as minister, a house was built for his occupancy; the site chosen was on the west side of Hallgate in front of the old Meeting House. There are both Day and Measured Accounts for this house in the Anelay Account Book (fols. 10, 245, 246 and 247). Work commenced in April 1803 with the demolition of a front wall and the digging of cellars. On 23rd of the same month a payment is recorded by Mr. Hodgson of £5.5.0 on account, a fact which enables us to identify the 'Chapel House' as belonging to the old Presbyterian connection.

The final items in the Day Account occur in January and February, 1804, they are for white-washing and the making of winebins; an item of interest in the Measured Account is the payment of 'Freight and Lock Dues for 9 Boat Loads of Wall Stones, 9 chaldrons at 16/- pr—£7.4.0.' The total of the Day Account amounted to £24.8.1½ and that of the Measured Account to £90.10.0, in all £114.18.1½. These were not paid until 12th July, 1805. The house is still in existence, with an entrance archway in its front giving access to the chapel; the lower storey has been converted and is now occupied as a shop.

The next major work undertaken for the Corporation appears to have been the first alteration to the Grand Stand in 1804. The stand, designed by John Carr and built in 1778,[24] closely resembled the York stand designed in 1774 by the same architect.[25] There are both Measured and Day accounts for the work (fols. 262-3) in the Account Book which throw considerable light on the alterations made at this period; references to these alterations by previous writings being somewhat obscure to the 20th century reader. The description in the 'Day Accompt' leaves no doubt as to the extent of the work. The engraving of the York stand shows a tall building surrounded on three sides by a balustraded balcony, supported by arches and with rusticated features at intervals below the balcony level. The main elevation was pierced by seven round headed

26 *Everingham Park, Yorkshire. Considerable alterations and repairs carried out for the Duke of Norfolk including the installation of this new staircase by William II. Architect, Francis F. Johnson.*

27 (above left) *The Stable block at Bramham Park after conversion into two cottages by William Peter Anelay, 1970.*

28 (above right) 29 (below) *Newburgh Priory, the Sitting Room before and after restoration by William Peter Anelay, 1966. Architects, Ward, Ruddick and Ward.*

windows, the centre one being of 'Venetian' pattern, there being a window of a similar type in each of the end walls. The Doncaster stand, as originally built, was similar in many respects, except that the seven windows in front were identical, the 'Venetian' window being omitted. The architect's plan for the proposed extension included the provision of a wing at either end with a slight projection from the front elevation, each having a 'Venetian' window in both front and end faces, the extension of the balcony in front of, and at the end of the new wing, and the extension of the balustrade round the top of the stand to include the two wings.

The accounts enable operations to be followed as the work went on; the first item would appear to the layman to be singularly irrelevant, namely the building of a Smith's chimney! But it must be remembered that the mason's chisels soon became blunted and would require re-tempering and sharpening. This, no doubt, is the explanation. In a further item 'May 14 to 19 . . . Cutting Bricks at Sandall for the front', a surprisingly 'modern' outlook is apparent; long before the days of Time and Motion studies and Critical Path Analyses, the master builder is using his common sense. It has already been stated that the intended wings were to be lighted by 'Venetian' windows of which the centre light had an arched head. The construction of this would entail the cutting of bricks to the wedge form necessary to build the arch. It will be apparent that the builder was using foresight in cutting bricks at the Sandall kilns, thus ensuring that only the proper bricks required would be brought to the site; whereas if the arch bricks had been cut on the building site it is certain that many more bricks would have been necessary to allow for inevitable wastage in the cutting, with a consequent increase in transport charges. The account also reveals that the requisite sand for the building was obtained from the tract of common land, known as the Low Pasture (fol. 262) 'April 16 to 21. Baring and Getting Sand on the Low Pasture', but whether this choice of site for obtaining sand was due to the builder, or to the frugality of the Corporation steward, does not appear. The Day Accounts, which refer chiefly to preparatory and cutting out work for the masons, furnish a complete picture of the sequence of operations. 'Cutting a Way for the Architraves'; 'Taking out Windows', this must refer to the removal of the end windows in Carr's existing stand; 'Turning inner Arches', these were presumably the entrance arches, from the existing saloon to the projected wings.

It is evident that the builders were working against time, for the September Race Week was rapidly approaching: 'Sept 17 to 22 Cleaning Long Room & Sundry jobs'. Work on the stand was resumed at the close of the meeting, but it is possible that these further operations were entailed by a change of heart as to the means of access to the new wings, as this had been by means of arches made in the existing end walls of the original stand. It is more than possible that the architect, William Lindley, was responsible for the decision to take down the inner walls, substituting for them an architrave at each end supported by two Ionic columns; thus incorporating the new wings into the existing saloon. The above suggestion that the alterations were made at Mr. Lindley's instigation is supported by the following entry, 'Oct. 19 to 24. pulling Down New Ceiling by Mr. Lindley's Order'. It is fortunate that this reference occurs in the Account Book, as it definitely ascribes the re-modelling of the stand to William Lindley, for entries in the Corporation accounts of payments to Lindley are often not specific, although it may be that Lindley's bills in the Doncaster Archives would provide confirmation. There is a contemporary account of the altered stand in the *Doncaster, Nottingham and Lincoln Gazette*, 20th September, 1805, but this contains no reference either to architect or builder.

The total amount of Thomas III's Day Account was £42.0.1¼, this payment occurring in the Borough accounts 'By Bill for Grand Stand in 1804'.[26] The Measured Account, as usual, gives little detail, but it is probable that it relates to the brickwork of the new wings, the total of this account being £77.18.11½, with a further sum of £13.15.4½ for plastering, this latter item would be for plain work, as is evidenced by the Accounts (fol. 262):

For What I did at the plastering	£12	5	4½
Allowance for the Lime and Hair that was sent to Rossington Rectory	£ 1	10	0
	13	15	4½

One, William Sadler of Doncaster, who was a plasterer was employed on the Grand Stand in May of 1805 when he received a payment of £10.5.0 on account 'By his Bill for Work done at Grand Stand in 1805 £67.12.4'. Sadler did not receive his final payment until 28th April, 1806, when he was paid the balance of £26.7.4.[27] On folio 263 is the Measured Account for the Stewards Stand, which was built by Anelay in 1805; the *Gazette* article mentioned above says' There also a new Stewards' Stand, built in an octagen form, and a separate box within it which entirely secludes the Judge from the company'. There is a reproduction of a colour print of the St. Leger of 1836 as the frontispiece to J. S. Fletcher's *History of the St. Leger Stakes* which shows this stand. Again, neither architect nor builder is named in the *Gazette* article, but the design may almost certainly be attributed to William Lindley.

The desire of the Rossington tenants to enclose their arable lands has already been mentioned, and their difficulties occasioned by the scattered portions of the glebe lands. Negotiations with the rector, ordered by the Doncaster Corporation in December 1800, did not reach fruition until 29th March, 1804, although it is evident that other considerations had emerged; on the above date the report of the committee stated that 'the rector of Rossington had agreed to advance £500 and such sum of money as arbitrators shall award for dilapidations, at the old rectory house and buildings, towards the expense of the new buildings and alterations at the new house according to Mr. Lindley's plans'.[28]

On 5th May, the Corporation petitioned the Archbishop of York (William Markham) to sanction their agreement with the rector. The petition says that the old house, which is in a low and unhealthy situation, is in decay and must be taken down, and is at a great distance from the parish church. After describing details of the intended exchange of lands, the petition further states that they will give the rector 'a new built house lately erected', adjoining to the churchyard, 'and to complete and furnish the inside work of the house and to erect new buildings for kitchens, stables, and all other out offices suitable to the said house, according to a plan drawn by Mr. Lindley, an able and experienced architect'. The petitioners further state that 'the cost of completing such buildings and finishing the inside of the house will amount to nearly £1,600, which with the sum of £800 expended in building the said house will make £2,400'.[29] To the casual reader, the account entries on fols. 266 and 267 must appear puzzling. They obviously relate to the one account, for the totals are carried forward to successive pages, yet the pages are headed as follows:—

folio 266 verso	Corporation Day Accompt for School House at Rossington
folio 266 recto	Continued
folio 267 verso	Corporation Day Accompt for School House Rossington Continued
folio 267 recto	Corp Acct for Rectury House Continued.

The solution to the problem appears to be that the recently built school house was the 'new built house, lately erected' referred to in the Corporation's petition to the Archbishop and which was destined (after addition and alteration) to become the new Rectory House; the builder was describing the house which hitherto he had always known (and built) as the School House, and it is only on the last page that he refers to it as the Rectory House. From the details in the Account, it is evident that the existing structure was undergoing considerable alteration:— 'pulling Down Walls Underpinning Stair-case Wall, Cleaning Bricks'; 'Digging Foundations for Passage & Studdy'; 'Getting Ground to Raise Laundry and Scullery'; 'Making Good about Frontispiece'. There are many other revealing details of alterations which must be omitted in this study. Thomas III's 'Day Accompt' for the Rectory amounted to £52.7.5¾ (fol. 267) and on 29th December, 1804, the builder closes the account with the following note: 'Given in to Mr. Lindley's up to this Date. Wrote out. Delivered Bill.', and in Mr. Anelay's account in the Corporation Ledger 'By Do. for the Rectory House at Rossington in 1804 £52.7.5¾'.[30] The Measured Account for the Rectory (fols. 269, 270) contains extraneous items such as 'At Blacksmiths walling'. When the charges for these are subtracted the total for the Rectory appears to be £73.17.3½. The rector, the Revd. Dr. Stovin, was the eldest son of James Stovin, Esq., a former Town Clerk of Doncaster, who resigned in 1779 and built a house called Shooters Hill, on the site of which stands the present Rossington Hall.[31]

There is no doubt that the worthy rector would be especially interested in the kitchens of his new rectory house, for he is reputed to have exclaimed upon one occasion that 'a goose is such a difficult bird, there is too much for one, and not enough for two'. That he was of generous proportions is born out by a contemporary diarist who, on 18th January, 1806, records that Francis Chantrey (later Sir Francis Chantrey, the sculptor) made a short stay in Doncaster with a view to obtaining portrait commissions. He says that Chantrey sent out applications to the local gentry which unfortunately could be read without unsealing, and continues 'one of which was sent to the Revd. Dr. Stovin, and which would have cost the *fat rector* of Rossington the immense sum of 2d, was returned unopened but the purport of it could not but be known to the holy man'.[32] Whether he could afford 2d. or no, the rector experienced some little difficulty in 'raising the wind' for the final instalment of the agreed sum of £900 for dilapidations and for his contribution to the cost of the new rectory house. The last entry in his account with the Corporation runs thus, '1807 Feb. 12 By a Note dated 9th July 1806 from R. Wilkinson Treasurer of the Turnpike from Doncaster to the south side of Tadcaster cross, payable to the Revd. Jas Stovin and Transferred by him so as to Balance this account £100'.[33]

It is apparent that both Thomas III and George Burks, the Corporation steward, were more regular in their business habits than their predecessors for no less than 45 payments on account were made to Anelay between 19th November, 1803, and 5th April, 1805. In the year in which the alterations to the Grand Stand were carried out (1804) a portrait of King George the Third was presented to the Corporation by Sampson, Lord Eardley, who had been made an honorary freeman of the borough, on 8th December, 1803. In the *Courtiers*, it is recorded '18 September 1804. Ordered that the Mayor write to Lord Eardley to thank him for his kind benevolence in making them a present of so rich and elegant a picture of our beloved sovereign, King George the Third'.[34] From the account book it is apparent that Thomas III was again working against the clock (fol. 256):—

30 to 8 Sept. Cutting Way for Picture in Mansion House

1 Day Self	getting Scaffold	3/6d

and after subsequent entries from the 10th to 15th September, the final one appears on 17th September:—

Finishing about Picture. 1 Day Plasterer. 1 Day Labourer 6/5d.

No doubt the City Fathers would be anxious to display their new treasure to those attending the balls and assemblies in the Mansion House during the coming Race Week. It will be noted that the picture was installed even before formal thanks to the donor were ordered. The picture hangs in the ball-room, and is surrounded by a plaster frame, crowned by a broken pediment supported on console brackets, but whether this is the original work carried out by Anelay and his plasterer cannot be said, but from the prices paid it would seem very doubtful.

In the ensuing year (1805) it was determined by the Corporation building committee that certain alterations should be made to the Mansion House, according to plans made by Mr. Lindley.[35] The Corporation Measured Account for the year 1805 (fol. 250-251) is again an 'omnibus' one, and apart from the numbers and daily rates of work people employed under the heading 'At Mansion House', gives us but little assistance. The main portion of the intended alterations was the provision of a new Dining Room, to be 60 feet long by 27 feet wide and 22 feet in height. A contemporary description mentions that there were 'three recesses for sideboards, at each end, and in the centre of the side opposite the windows, which give the room a very handsome appearance. The centre one is of elliptical form, very spacious and contains the orchestra which is decorated with four lofty columns of the Corinthian Order, supporting an ornamental frieze and cornice. The ceiling is coved and formed into various compartments, three of which are ornamented with vines, festoons &c from which are suspended three glass chandeliers. This room with the additions and improvements lately made at the Mansion House, are from the designs of Mr. Lindley, architect'.[36]

The period over which the payments to Thomas III were made extends from 8th July, 1805, to 22nd February, 1806, and it is interesting to find that payment was made to Messrs. Rickard 'By Bills for Mansion House &c up to February 15th 1806—£96.12.0½'.[37] Moreover, there appears to have been the customary jollification (at corporate expense), for the workmen employed, for William Dodsworth of the 'Green Dragon Inn' submitted his account '1806 February 28. By a Bill for a Rearing Supper'.[38] It is submitted that the above coincidence of dates leaves little room for doubt that the Anelay account book entries relate to the building of the new Dining Room, for which the builder charged £101.13.10¾. The Corporation Ledger also records (fol. 124) that Mr. Lilley of Doncaster, on 7th October, 1806, sent in his account 'By a Bill for 3 Side Board Tables &c for the Mansion House in 1806—£49.8.6'. A further item paid to inn-keeper James Farrington '1806 Aug. 29 By Dinner Bill on fixing up Chandelairs. £1.17.0'.[39]

Among other works carried out by Thomas III at this period (actually in 1808) was the brickwork of the School of Industry. This institution for the lodging and training of young girls of the poorer classes had originally been housed in an old Tudor building to the west of the church, but with the imminent sale and demolition of the property, consequent upon the death of the owner, it became necessary to provide other accommodation for the charity. A site was chosen to the north of the dispensary, and bordering upon the Great North Road. Here a new school was erected; Thomas III's payment for the work was over £50, but the precise sum is not clear from the account, as two totals are shown. The Corporation Ledger contains an entry of a payment to Messrs. William and John Rickard of Doncaster, for the carpenters' work of the building 'To a Bill for Measured Work for the School of Industry—£155.5.6¼'.[40] In the year 1816, on the temporary suspension of the charity, the house was let to private tenants, and survived until after the building of the North Bridge in 1910-11.

A commission undertaken in the year 1814 is not without interest. It appears on fols. 296/297 and was for the building of bridges on the Potteric Carr, a recently drained area to the south of Doncaster. Considering the nature of the ground on which the bridges were built, it is curious that the work was carried on in what was the worst season of the year. Operations commenced on 19th September, 1814, and continued intermittently until the 23rd March in the year 1816, payments being made to the builders at approximately weekly intervals. It will be apparent that the bridges were small occupation bridges over drains with which the district abounded. The bricklayers were paid 4/- a day instead of the standard rate of 3/6 while the labourer received 3/- a day, or 3d. above the standard rate. It is suggested that this increment may have been paid owing to the unpleasant working conditions.

In the year 1814 Mr. Fitzgerald, the lessee of the Theatre, made representations to the Corporation as to certain improvements to the building, whereupon they decided to ask that plans and estimates be submitted.[41] From the context of a subsequent order it is apparent that the architect consulted was Lindley, whose firm (he was now associated with John Woodhead and William Hurst, jun.) lost no time in the preparation of the required plans, and it was resolved on 7th December, 1814, that they be carried out.[42]

In the Anelay account book there is a measured account (fol. 295) for work on the theatre, again very meagre in detail. Included in the work was the making of a new drain, beginning in New Street near the Theatre and ending at the White Bear in Hallgate. The course of this drain is indicated on a loose sheet inserted in the back cover of the account book; being a note of Mr. Lockwood's certification of measurement which gives the course as 'New St., Scott Lane, High St. 519 yards 1 ft.'. It may be that the alterations to the Theatre had made new drainage arrangements necessary. The period over which account book entries are recorded extends from 27th March, 1815, to 31st June, in the same year, the amount being £56.2.0.

The Corporation Cash Book contains payment for 21 chaldron of wall stone to the theatre and also eight entries relating to the sawing of over 1,600 feet of timber for the same purpose, the period covered being from 14th March to 8th July, 1815.[43] There is a full account of the appearance of the altered Theatre in the *Doncaster, Nottingham and Lincoln Gazette* (22nd September, 1815), which is too long to include here, but the chief alterations made at this time included the building of a semi-circular extension at the rear of the stage (or the east end) which contained the green room and ladies' dressing rooms, which hitherto had occupied part of the stage area; the provision of new staircases, and the removal of the principal entrance from the north side to the west end where an entrance was made through a new Doric portico, surmounted by a balustrade.[44]

Steps were taken in the year 1823 to provide more permanent premises for the publicans of the town, who had hitherto sold liquor on the race-course from a line of tents, the sale of liquor being forbidden in the low rooms of the Grand Stand, in November 1793.[45] The 42 booths planned were commenced, according to Mr. Anelay's Measured Account, on 5th May, 1823 (fol. 298). Whether a mistake had been made in the plans, or whether the architect had changed his mind, is not obvious, but the account (fol. 297) 12th to 17th May, has an entry 'pulling up Foundations that was laid by Mr. Hurst's Order. 5/10'. The Mr. Hurst referred to was William Hurst, a partner in the Doncaster firm of architects Messrs. Woodhead and Hurst, so we may conclude that they were responsible for the planning and design of the booths. Work on a stone wall and upon the booths continued until 13th September, 1823. The booths were probably completed at that date, but a small amount of work was carried out on the stone wall in the months of October and December.

The new work evidently adjoined other buildings in the locality, for several items in the Day Account relate among other things to 'Repairing Mr. Hawley's fireplace'; 'Building a wall next Mr. Hawley's Wash-house' (fol. 297). It is possible that old buildings were connected with the Grand Stand Keeper's premises, for Mrs. Mary Hawley is referred to as keeper in 1837. Other work particularized included 'pulling down Stone Wall Next Rouber House for the Chimney'; 'putting on Chimney pott on the Stand'; and 'Colouring part of back wall next the Field by Mr. Hurst's Order'. The 'Rouber House' must refer to the rubbing house, where horses were rubbed down after the race, and which was an erection contemporary with the original stand. In the Doncaster Museum and Art Gallery is the first Rent Book of the Race-Course Booths 1823-1869. As above stated, 42 booths were built in 1823 and a further six in the year 1851.[47]

On 2nd January, 1824, the *Doncaster, Nottingham and Lincoln Gazette* announced that it was the intention of the Corporation to provide further accommodation on the Grand Stand; the journal says that 'they had fixed upon a plan of enlargement, which will increase its capacity to accommodate one thousand additional persons'. Briefly, the alterations comprised the addition of a cast metal verandah, on the roof of which were a series of steps; the provision of more steps on the top of the stand, and the building of a semi-circular extension to each end of the lower balcony, with the necessary additions to its supporting colonnade. These alterations were made to the designs of Messrs. Woodhead and Hurst, architects, who were the successors of William Lindley who had died in the year 1819.

In addition to the above detailed alterations, considerable internal re-construction was entailed such as the provision of staircases, vestibules, &c. The Anelay account book (fol. 300) informs us that work had commenced on the Stand on 2nd February, with the digging of foundations. This is the only detail given in the Measured Account, except for the significant memoranda (5th to 10th April) to the effect that wages had been advanced; the bricklayers daily rate being increased from 3/6 to 4/-, while the wage of labourers had risen from 2/4 to 2/8 per day. Building work seems to have commenced in the middle of March and carried on to approximately the first week in July, when plastering of the new building commenced. This work was probably completed on or about the 11th September, but the colouring of the exterior (also paid by measure) had begun in the last week of August. Apparently the builder only supplied the labour for this latter work, which must have been hard on the brushes for there is a charge 'To cost of Brushes to Ditto 9/6' (fol. 302). The Measured Account for the Brickwork amounted to £82.6.2 and the plastering cost £17.16.2, making a total of £100.2.4. The *Gazette* of 20th August, 1824, waxes eloquent about the nearly completed work, 'The improvements in our Grand Stand are drawing fast towards completion, and we may safely aver that when finished it will surpass any edifice of its kind in all Europe'.

The Court accommodation provided in the old Town Hall had long been the subject of complaint by the County and Borough Justices, and not without reason. To remedy this the Corporation entered upon a scheme to improve the Hall late in the year 1828. 'The Corporation are also about to make considerable alterations and improvements in their Guild Hall and Gaol, both having been found extremely inconvenient and insufficient for the purposes of the borough'.[48]

The alterations thus envisaged were carried out by Thomas III in concert with another local builder, John Reasbeck, and on fol. 317 details of this arrangement appear. 'Contract for Altering Town Hall between Reasbeck and Anelay sen^{r} . . . . . My Contract £30.10.' The builders were engaged on the work from 17th November to 20th December, but this is not the sum total of the work carried out by Anelay, for on fol. 129 there is a parallel account for 'Extra work at the Town Hall'. This extra work included alteration to a staircase, namely walling, pricking-up and repairing flags, also the provision of a new chimney pot. A further item would seem to have been the conversion of an existing shop into a prison. At first examination it appeared that the shop abutted on the hall, but on further investigation it seemed that the shop was actually part of the lower storey of the hall, as in a report of the Corporation Commission in the *Gazette* of 20th December, 1833, the cells are stated to have been under the town hall.[49] The account book details, 'pulling down walls in Mr. Nicholson's Shop . . . . . Building walls for to make Prisons &C'. It is possible that these cells were constructed at this time as temporary accommodation for evil-doers, while the projected new Gaol was under construction, for in the true spirit of economy it was ordained that some of the materials from the old Gaol should be used in building the new one, an intention which was carried into effect. In addition to this work, the court-room in the Town Hall was also altered, an improvement which elicited complimentary remarks from the Grand Jury, whose foreman stated that he had been requested to 'signify their extreme satisfaction in witnessing the alterations and great improvement which had taken place in the Court-House since they last assembled in it'. Further comment referred to the fact that 'it was executed with great professional skill and effect'. The two accounts quoted furnish the commencing date of 17th November, 1828, while small finishing work, such as the whitewashing of the cells and pointing slates in part of the Town Hall roof, were being carried out on 3rd January, 1829.

Early in the year 1831 the Corporation were of the opinion that the Mansion House was not large enough for their civic entertainments, and on 11th February passed the following resolution: 'For rendering sufficient accommodation to the Nobility and Gentry attending Doncaster Races, and for improvements and additions to the House, Resolved that Mr. Hurst's offer to carry the whole into effect for £1,500, including every expence according to the Plans now produced, be accepted.'[50]

As in the case of the Gaol, which is previously mentioned[51], Alderman Hurst's sanguine estimate of £1,500 was greatly exceeded, the additions costing some £4,000, The alterations and additions made at this time included the provision of a saloon with folding doors communicating with the Ball-room and Banqueting Chamber; household offices in the newly opened Priory Place, which comprised a new housekeeper's room, butler's pantry and store room; the ceiling of the card room was also raised. A large mirror was purchased in London and 'placed in the Recess of the supper room to reflect the Vista of the suite apartments.'[52] Further expensive items were two chandeliers, one for the new saloon and the other for the card-room.

The price paid for this magnificence is recorded in the Corporation Accounts:—

Sep 3rd (1831)	British Plate Glass Comp^{y} for Glass for Supper Room	£158.0.0.
„ „ „	Mr. Skidmore for Glass Chandeliers for the Saloon	£131.5.0.

also recorded is the price paid for the site of the new domestic building:—

1831 July 4th. Mr Ald. Sheardown for the purchase of Ground in Priory St. £306.0.0.[53]

Mr. Anelay can hardly be blamed for the excess expenditure as his Measured Account with Messrs. Woodhead and Hurst did not amount to £30. The account again is not informative, but it is seen that Thomas III, when working as a labourer only rates himself as such, and also that while William Hurst may have handled the talking department of his firm, it is noticeable that his senior partner, Mr. John Woodhead, kept his hand on the purse strings, the payments on account all being made by him.

It is hoped that this short account of some of the building transactions of Messrs. Anelays during an early period of the firm's existence may not be without interest.

The Building Trade in the Early Nineteenth Century

The following comments may serve to throw some further light on the nature of the work undertaken, on the practices of the building trade, and on the materials used at this period. At the outset it should be noted that rigid economy was exercised in the recovery and re-use of building materials. There are several instances in the account book of the dressing and cleaning of bricks saved from a demolished structure, with a view to their subsequent re-use, namely at the Grand Stand (fols. 264 and 265) and at the Rectory House at Rossington (fol. 266). The same may be said of materials in which lime was a constituent. Lime scraps were riddled for further use (fol. 207) 'March 8 to 13 . . . . . Ridling Lime Scraps . . . . .'; in this account they were used for levelling the floor, but in some instances the scraps were beaten, presumably the end product being re-used in coarse lime mixtures (fol. 266).

The above examples are dated 1804, but evidence of this practice occurs earlier, in the year 1798 (fol. 95), 'Mrs. Ruth Smiths Day Accompt . . . . . 26 and 27 Oct. Removing and Beating Lime and Hair from Butter Cross . . . . .' It is apparent that the beaten lime and hair was destined to be used in the repair of Mrs. Smith's house. A further point of interest in the above connection is the revelation that the Butter Cross had a lath and plaster ceiling, which was renewed by Thomas Anelay III in the year 1804 (fol. 255).

In the account for the Grand Stand mentioned above, there is an item which refers to the provision of scaffolding and the making of mortar for the masons. This implies that it was in the province of the builder to provide both scaffolding and mortar, even though the masons using them were employed by a masonry contractor. No doubt there were sound economic reasons for this arrangement, as the builder would require scaffolding for his own operations, and the amount of mortar required by the masons would be comparatively small.

The turn of the century marked innovations in the furniture of the domestic hearth. Up to this period the majority of houses had open hearths, the cooking pots being suspended from chimney cranes, or idle backs, or alternatively placed on stands which stood on the hearth over the fire. In larger houses these arrangements were supplemented by the roasting-spit fixed in front of the hearth, the revolving of which was performed by various means, among them being the weight driven jack and the smoke jack. This latter, as its name implies, was operated by the hot gases ascending the chimney flue, the jack being mounted over the lintel of the fireplace, the driving shaft passing through the flue being fitted with vanes on which the hot air impinged, thus revolving the shaft. On the outside of the fireplace a long shaft was operated by the jack; this had pulleys at either end from which belts descended to pulleys on the spit.

An earlier reference (17th May, 1790) to the intended fitting of a 'Smoak Jack' in Widow Sykes' house at Rossington, is described in the Corporation accounts[54] where Thomas II and J. Britton, the whitesmith, accompanied the Corporation steward to examine the widow's chimney in order to give their opinion as to the feasibility of installing a smoke jack. There are several references either to the fitting, repair or alteration of smoke jacks in the Anelay account book; folio 43 (Mr. Tomlinson's account) refers to the putting up of 'Smook jack' in 1798, and in Miss Murphy's account (fol. 229) the setting of a 'Range Oven' seems to have entailed 'Altering Smoak jack'.

The years between 1790 and 1805 saw the introduction of the cast-iron range and oven. One of the first was patented by Thomas Robinson in 1780; another, the Bodley Range, was patented in 1802. This latter was an advance on the Robinson range, having flues on three sides of the oven, the top of the grate being enclosed, and the front of the grate utilized for roasting. This latter was the direct ancestor of the later kitchener ranges.

The many entries in the accounts over this period for setting ranges is thus explained, although Thomas II was apt to describe a simple fire grate as a range, or vice versa, a habit which renders an exact interpretation difficult. For instance, on folio 146 (Corporation Day Account), the entry reads: 'April 28 to 3 May 1800 . . . . . Setting Range in Council Room', this would almost certainly be a fire grate. The first example which we could definitely describe as a range occurs in the account of Richard Ainley [sic] where the entry runs 'Feb 21. 1800. Setting range and oven . . . . .' It would appear that these ranges installed by Anelay at this early period were of the 'Robinson' type.

Another early 19th century cooking range was the invention of Count von Rumford, who had carried out many experiments on the economic utilization of heat. The Count (1753-1814), born in Colonial America, was a Bavarian statesman and general who subsequently became a British officer—but above all he was a great physicist and investigator of latent heat. His ranges were scientifically planned, often being of horse-shoe form so that the cook could have equal access to all his saucepans, which were sunk in holes below the top surface of the range. The ranges would seem to have been constructed in brick with an elaborate flue system. When Rumford came back to England (c. 1796) he wrote: 'The kitchen fireplace of a family in easy circumstances in this country consists almost uniformly of a long grate for burning coal placed in a wide and open chimney'. He was an advocate of small iron ovens (most of them at this period being constructed of brick), and before the year 1800 had designed 'small ovens for poor families'.[55] We have no knowledge of the exact form of the Rumford range which Thomas III constructed for his wealthier clients, or how he became possessed of the knowledge which enabled him to build them. It may be that plans had been obtained by one of his patrons and which the builder afterwards turned to good account. There are several references in the account book 'to Rumfordising grates'. The first grate of this pattern to be made by the builder was for Lord Stourton who lived at Stapleton Park, near Darrington, in the West Riding of Yorkshire; from 'Lord Stourtons Day Accompt' (fol. 116) we gain some slight idea of the work involved.

It would appear that Thomas III rode to Stapleton in March 1798. At this time he had no horse and paid 3/- for the horse-hire. It is probable that this first visit was exploratory, for nothing further emerges until the June of the following year, when he again visited Stapleton. The following week the builder's men are working at the range, or as Anelay describes it, 'Rumfordising Grates'. A bricklayer and mason were engaged for approximately nine days, during which time 300 bricks were used. At this period the builder again went to Stapleton, probably to satisfy himself as to the progress of the work. The total of the builder's account for the work amounted to £7.12.0, but of this £4.4.0 was for 'a Bright Wrought Iron Gate', so that the cost of the 'Rumford Grate' was £3.8.0, although this did not include either the mason's wages or the cost of the bricks (fol. 116).

Several like conversions appear in the account book, among them the installation of a 'Rumford Grate' in the Doncaster Mansion House in the September of 1801 (fol. 149). The first and only mention of the employment of iron occurs on folio 312 'Mr. Tinklers Day Accompt for the Falcon Inn', where appears [Mr. Anelay's men] 'Assisting Foundrymen with Iron Joists . . . . '

There are interesting references to the use of clamp bricks on the several folios (292, 301, 302, 305, 306) of Captain Ramsden's Account (1842) for the building of a house at Hexthorpe. The first two entries (fol. 292) record the purchase of these bricks from local brickyards, but those on the remaining folios are described as 'from the Canal'; 'from the Cut'; from the New Cutting', etc. These latter phrases gave rise to a certain puzzlement as to their meaning, but if it is recollected that a new canal cut had been made at Doncaster at this period, through the low-lying common ings land known as the Crimpsall, the conclusion could be that the excavated warp had been used to make clamp bricks. Warp was commonly used in the brickyards of the Isle of Axholme.

As clamp bricks have been mentioned in the foregoing paragraph, it may be desirable to say what they were. They were burned in systematic stacks, or clamps; no kiln was required as the requisite fuel was incorporated in the body of the brick when it was moulded, more fuel being sprinkled between the layers when the bricks were stacked. Fires were then started in channels made for that purpose, which then ignited the fuel between the layers and also that in the brick itself. The fire when once started in the lower courses, required no further stoking, the heat spreading gradually through the entire mass. It may be of interest to observe that 'Old Brick Kilns' are marked on the O.S. map of 1890, in the Marshgate locality, but hitherto there has been no reference to the possibility that the brick material was dug from the canal.

Another innovation in which the services of Thomas III were sought was the installation of water closets; although the invention of such a convenience was attributed to Sir John Harington in 1595, few houses were so equipped until the beginning of the 19th century. Probably the first introduction of the water closet in Doncaster was in a house, now No. 18 South Parade, about the year 1798, although this statement is subject to correction. The absence of a general sewage system in towns and cities militated against the general introduction of water closets. It is pertinent to say that the sewage system of London was not made until the year 1865, and even after the provision of adequate works, many earth privies still survived. There were still 'necessaries' of this type in Doncaster till the second decade of the present century. In 'Miss Murfey's Day Accompt' (fols. 229 and 253) are the details of the building of a water closet with its attendant reservoir or cesspool. The entries extend from 2nd November to 26th December, 1807. Miss Murphy conducted an academy for young ladies in premises on the west side of South Parade, known as Hall Cross House, which was demolished in the early years of this century. Evidently the builder thought that this commission might be the first of many, for at the foot of the account he saw fit to abstract and record the price of the work 'Cost of Water Closet £13.1.1¼'.

In 1823, Thomas III, in company with his sons Thomas IV and Charles, made 'a New Drain on the Common and a reservoir in Mr. Hawley's garden'. As the latter was the Keeper of the Grand Stand at this date, it is suggested that these works were made in connection with the installation of water closets in the Grand Stand, but of any actual installing of such conveniences no details are given. From this same account (fol. 300) the information is gathered that Thomas IV, at the age of 17, was rated at 3/-, whilst Charles, some two years his junior, received 1/6, the full rate for a bricklayer at this time being 3/6.

Besides his building work, the services of Thomas Anelay III were at times called upon to value property. This he undertook for Mr. Sheardown in the year 1811 (fol. 142). 'To valuing Mr. Sheardown's property in Pell's Close'. For this he charged the modest sum of 10/6.

In his account with Mr. Pollard, Plasterer,[56] (fol. 137) he charges 10/6 for 'Measuring and Making out the Account of Mr. Dale, white-washing and colouring at Warmsworth'. Besides the actual valuing it would seem that it was the business of the valuer to make out the tradesman's account for the presentation to the customer, A further account (fol. 124) reveals that it was the accepted practice for the measurer to charge for his services at a rate of so much in the pound on the total of the account. In this case the rate was 2½ per cent, or 6d. in the pound.

These are all very comfortable activities, but several account entries tell us that the builder and his men occasionally experienced moments of real danger. One of these times of crisis occurred on the night of 21st January, 1802. On this night, according to the *Doncaster, Nottingham and Lincoln Gazette* of 22nd January, roofs were stripped of their tiling or slating, and chimneys thrown down in a great storm. The builder's description is less dramatic; he writes: '1802 Jan. 21. Securing the slating in the Storm . . . . . 0. 5. 0.' (fol. 85).

Another episode which must have been somewhat uncomfortable for the bricklayer involved is recorded somewhat prosaically in 'Mr. Heaton's Day Accompt . . . . . 1806. 23 to 28 June. Catching Bees 3/4 Thrush 0. 2. 7½.' It is probable that the bees had found a home in one of the roofs of the Parsonage premises and had to be disposed of before repairs could be effected (fol. 183).

A few words must now be devoted to the building materials of the period as described in the account book. Possibly the principal of these was lime. The use of cement is not recorded very frequently; the first instance of its use seen hitherto appears in Mr. Tinkler's Day Account (fol. 313) when 2 pecks were used in repairs to a malt cistern. Its use is recorded again in Captain Ramsden's Account for the building of a house at Hexthorpe, dated May 1842 (fol. 292): it cost 26/- per barrel. Portland cement was allegedly invented by Joseph Aspdin in the year 1824. Lime entered into many building compounds; with sharp sand it formed the mortar of the period; when mixed with either horse or cow hair in varying proportions, it was used for many purposes and there are countless references to its provision and use in the account book. Lime-hair mixed with a certain quantity of sand was used as the first keying, or 'pricking up' coat in plastering. In this form it was referred to by the plasterer as 'coarse stuff'. Further instances of the use of lime-hair are the interior pointing of pan-tiles; the setting of firegrates; the fixing of chimney pots; and in conjunction with pebbles for pebble-dashing the exterior of buildings.

The 'fine-stuff' used by the plasterer was made of well slaked lime; for some purposes a small quantity of hair was used in the mixture. Another use of fine lime was in the preparation of 'putty' (not to be confused with the putty of the present day). Here the lime was mixed with water so completely as to be held in suspension and was of the consistency of paste; this mixture was then allowed to stand and evaporate until it was of a suitable working consistency. Lime putty was used in the building of a superior brickwork called 'gauge' work, where it was employed in the place of mortar. It is mentioned in the book (fol. 28) as a basic material used in white-washing. A further reference (fol. 130) implies that when used for this purpose it was mixed with size; it was often the medium for applying colour (fol. 135) 'To fine lime To Mix to Colour . . . . . 0. 6. 0.' From folio 134 it is gathered that the cost of lime putty in the year 1801 was 6d. a pail, the same price being entered in 1830.

The builder obtained his bricks from local sources. Those used in the town would doubtless be got from the kilns at Sandall, which was owned by Alderman Thomas Jennings in 1773, and then by Mr. Lockwood, from whom they passed to Alderman Mandall (fol. 292). Bricks required for use in and about Rossington came from the Corporation kilns at that place, which have previously been referred to. The builder's sources for lime and walling stone have also been mentioned. The purchase of bricks from Misson on the boundary of the Isle of Axholme is also recorded in Mr. Heaton's Account (fol. 178).

The hair used in the manufacture of lime-hair was obtained from local tanners, the tan-yards being situated in the Fishergate locality near the River Don; hair was usually sold by the mett, which in Yorkshire appears to have been the equivalent of two bushels. One of the tan-yards, occupied by the Quaker family of Clark, was in production in the early years of the present century. In Mr. Anelay's Contract with Mr. Reasbeck (fol. 317) is the following '1828. 15 to 20 Decr. To 3 Metts of Hair from Clarks . . . . . 7. 6.'

Reeds were purchased by Thomas III, both for the keying of ceilings and of plaster floors. There is an entry in the account book recording his purchase of 300 bunches from a person in Lincoln. Considering the proximity of marshy tracts in the Isle of Axholme, which one would suppose would yield reeds in abundance, it is surprising that the builder had to go so far afield for supplies of this commodity. In 1807 (31st August to 5th September, fol. 177) the builder sold to Mr. Richard Tootle '40 Bunches of Reeds at 4d pr Bunch'. Mr. Tootle was a keel-builder, and it is possible the reeds were for use in his craft, but of this there is no certainty.

The measures used in the accounts for charging for mortar and lime-hair were the 'board' and the 'barrow', the same measures being used by Thomas II in the smaller day-book (1772). In comparatively few instances is mortar charged by the hod; which was a measure of quantity quoted in builders' price books; from the prices quoted, however, the following relationship emerges:—

1 Board	.....	½ a hod.
1 Barrow	.....	3 Boards.
2 Boards	.....	1 hod.

The above relationship is sufficiently close for a comparison to be made between the Anelay Account Book and *Crosby's Book of Builders Prices, 1804*. The prices quoted in the latter are presumably those prevailing in London and the south of England. These are given in the appended tables.

There is one question relating to the Day Accounts for which no answer has been found. Where does the builder's profit margin appear? The amounts entered in Anelay's book as a total also appear in the Corporation accounts as being paid to him, yet the details entered by him only include the wages of his staff and the cost of materials used.

Early Nineteenth Century Doncaster: the Domestic and Economic Scene portrayed in the Accounts

It is fitting that this starts with Thomas III, an honest soul, a man who had no false pride and who did not regard labouring tasks as beneath his dignity; there are several accounts which contain the entries 'Self as Labourer', and in each case the hours worked are charged at the labouring rate.

It also emerges that he let lodgings, the book containing accounts of three tenants, extending over a period of approximately 14 years, from 1817 to 1831. Perhaps he imagined that this additional source of income would be useful during slack periods in his trade, but if he cherished such notions he was doomed to disappointment, for his several tenants were often more than six months in arrears.

The first tenant (fol 156) was Thomas Whitall Waterworth, a local solicitor and son of Alderman Thomas Waterworth, who was in business as a carver and statuary, and who supplied the 'Emblem of Health' for the Dispensary, built by Thomas II and George Hewitt. The account does not reveal when Mr. Waterworth relinquished his lodgings, but the settling clause is of interest, for besides the statement that Mr. Anelay's tenant had paid up to the 1st October, we see that the builder had money at interest '—also the Interest of £100 Now in Mr. Robert Mittin's Hands at 5 pr cent up to the 15 July 1817. I say Received'.

The 'Robert Mittin' was Robert Mitton, stated in the Doncaster Marriage Register to be 'of Hooton Pagnell', where he married Mary, Mr. Waterworth's sister, by licence on 1st November, 1813.[57] A Robert Mitton is listed in the Directory of 1838, where he occurs as a farmer in Badsworth:[58] Dr. J. Sykes' pedigree of the Waterworths also refers to Mitton, as of the latter place.[59]

The second tenant was a Miss Sidall, who lodged with the Anelays in 1822 (fol. 54). She was evidently in the drapery business, for the worthy builder did a little owing on his own account, or rather on Miss Sidall's, 'Self Dr for Cloath and Different things for My Son to be paid for at Different Installments'. From 8th June to 9th November, 1822, 13 amounts are entered, totalling £2.16.6d. in payment of the builder's debt, but his lodger did not pay up until 1st January, 1823; 'To 30 Weeks Lodgings from the 8 of June to the 4 of Jany 1823 at 3/- pr Week Comes to . . . . . 4.10.0': and 'Setteled with Miss Sidall up to the 1 of Jany 1823'.

At this period there appears to have been a depression in trade (see Graph, page 60). The prices of live-stock had fallen by two-thirds and wages had suffered a decrease. It is worthy of note that the account does not contain a single entry for the year 1822, and while we do not suggest that Thomas III was on the verge of bankruptcy, it is apparent that he must have had a difficult year. This may partially explain his 'Installment' agreement with Miss Sidall.

After the lady's departure another attorney, Mr. Pearson, moved in, and he would seem to have become a family institution for he resided with Anelays for over six years, from 13th December, 1824, to 12th July, 1831.

It is possible that the need to husband every shilling induced the builder to let part of his premises to a near neighbour, for on the 12th June, 1821, Anelay records, 'Mr. Thirlwall Entered to the Rent of my Shop Down My Yard at the Rate of 1/- per Week'. Mr. Robert Thirlwall was an Ironmonger and Hardware Dealer, and had a shop in the Magdalens. But if Thomas III expected the prompt and regular payment of his shilling he was sadly mistaken, for by the 22nd September, 1823, he had only received about 22/-, although he endeavoured to redress the balance by obtaining several items of hardware from Mr. Thirlwall's stock; these included padlocks, three rules, a 'Garden Hoe and Rake', a frying pan, a shovel, and knives and forks.

But despite his efforts, Thomas III could only sadly record at the beginning of 1823 'To 81 Week Rent Due the 1st January 1823 at 1/- pr Week—4. 1. 0.' Of this debt, he had only recovered some 18/- by the ensuing September, and had to wait for the balance until February 1828;

thereafter Mr. Thirlwall's payments were frequent up to the 13th March, 1832, when the account ends (fol. 159).

In earlier and more prosperous days (at least for the builder), there occurs in 'Mr. Hepper's Day Accompt' (fol. 202) the following: 'April 7. 1804, To 5 Silk Handkerchiefs at 8/6 pr . . . . . 2. 2. 6'. There is here no suggestion of 'Installments'. Mr. Hepper was a Hatter and Hosier, with a shop in the High Street; a further transaction with him appears in a folded bill, placed in the front inside cover of the account book. It reads:—

25

Mr. T Ainley   Doncaster.

1814	Bot of S. Hepper	£	s	d
Jany 4th	1 Pr Woms Blk Hose		3	0
April 26	2 Pr Woms Do		5	0
Sept 1	2 Pr Blk Do		5	0
,,	2 Pr Maids Wh Cotton		4	6
Dec 12	1 Youths Black Plate		12	10
,, 27	To Old Boots	1	6	0
		2	16	4

What, may be asked, were Thomas III's motives in the buying of old boots. We can only surmise that their ultimate destination, like the spetches mentioned later, was the glue pot.

The builder's men were evidently expected to turn their hands to any job which came along. An entry in 'Mr Wroughton's Day Accompt' (fol 130) runs: '1799. Feby 14. Assisting Down with the wine 5/0 . . . . . 0. 5. 0'. This Mr. Wroughton owned Adwick Hall, but had recently acquired a house in Doncaster (now No. 16 South Parade) and although the house had been but recently built, Thomas III had been employed in carrying out some internal alterations, in addition to the building of dog kennels, garden walls, dairy, larder, etc. From this same account is derived the information that the builder sometimes employed casual labour: 'Aug 4 to 9. Paid a soldier for Five Days . . . . . 11/3'.

A further instance of the employment of Anelay's men in work, other than their usual avocation, occurs in the Day Account of Mr. Thomas Hall (fol. 308): '1833. 5 to 10 July. At Harvest Work, Do Assisting the Carpenters with the Roof &c. 3 Days Bricklay^{r} 3/6. Labourer 2 Do 1/6 . . . . . 13/6'; and on '12 to 17. At Harvest Work, Making piggery &c. 9 Do ½ bricklay^{r} 3/6. Labourer 6 Do 1/6 . . . . . 2 .2. 3'.

Mr. Thomas Hall had a farm, referred to in the account as St. James. This was situated near the present St. James, or Hexthorpe Bridge, and was swallowed up in the building of the Great Northern Railway; in addition to being a farmer, Mr. Hall was also a butcher, and in the above account there are two items relating to Mr. Anelay's purchasing of meat (fol. 294). 'Nov 24 to 29. 1821. 2 pounds of Beef Stakes, ... 1/4', and on the same dates, 'Beef . . . . . 1/4'. On folio 308, in the same account, there are two entries which relate to the selling of lime: 'Oct 11 To 1 Barrow of lime for the Wheat . . . . . 8d'; and on 'Oct 16 1 Do for the Wheat . . . . . 8d'. It is suggested that the lime was probably intended for use as a fertilizer.

There is another transaction relating to the buying of meat in the account of Mr. Bartholomew Atkinson for plastering, although here it would seem that Mr. Atkinson's ability to pay was suspect, and the meat was taken in lieu of money (fol. 74): 'Decr 16. 1800. To Half a pig of Bart Atkinson, 9 st. 2 lb, at 9s pr Stone . . . . . 4. 1. 0'. The full total of the account amounted to £13.12.4, on which Atkinson had previously paid £6.6.0. Whether the builder ever received the balance does not appear. It is possible that he, Atkinson, was the person listed as a plumber and glazier in the Directory of 1822.[60]

Another of the firm's customers was Giles Elliott (fol. 88), an ironmonger who lived on the north side of St. Sepulchre Gate, near its junction with High Street and Frenchgate, in premises belonging to Mr. Francis Turner. It may be of interest to state that he was a brother of Ebenezer Elliott, the Corn Law Rhymer. He did not long survive his establishment in Doncaster; he died 17th October, 1814, his widow Amelia carrying on the business. A note in the account runs: 'Trustees to pay the Ballance'.

There is a rather matter of fact entry, 'Novb, 1800. Setting Rang and Oven Where Teare Lived . . . . . '. Mr. Edward Teare had been an army surgeon, but practiced in Doncaster for a number of years; he was decidedly eccentric. The first notice we have of him occurs in the *Doncaster Journal* (12th April, 1788) in a paragraph advertising '6 Duetts for Two German Flutes', composed by Mr. Teare and published by subscription.

In the year 1797 Mr. Teare was again in the public eye, for on 8th December the above-mentioned journal carried a notice to the effect that Mr. Teare, Army Surgeon and Practitioner in Doncaster, had published by subscription, *A Treatise on the Use and Abuse of Tobacco*, price 1/6. Some two years earlier he appears to have loved and lost, if the following notice which appeared in the *Doncaster Journal* of 2nd May, 1795, is to be believed:—

(We are authorised and desired to insert the following)

Saturday, April 25th, 1795.

'The morning lours and heavily in clouds bring the day
The great, the important day, big with the fate of the bucket
and of broom'
Pictures and furniture in Neddy Teare's back room.

A SALE

On Saturday last was held a Levee, at Mr TEARE'S residence (not in Downing-street) but in the Market-place Doncaster; where there was such an assemblage of Vulture-Harpies, as is believed never graced PANDEMONIUM!

* * * The cause of this extra-ordinary visit has originated from the cruelty of a beautiful young Lady—and the villainy of a swindling scoundrel.

E. TEARE.

Surgeon to his Majesty's
late 85th Foot.

Of Mr. Teare's professional life in Doncaster, little is known. An entry in the Corporation Cash Book reads: 'Mar. 4. 1795 Mr. Teare for Curing Ed Mitchel's Shoulder . . . . . 10/6'.[61] Edward Mitchell was a corporation lamp-lighter, injured whilst attending to his duties and to whom a small pension of 3/6 a week was subsequently paid. Edward Teare died at the age of 42 and was buried on 6th April, 1799.[62]

It had long been the custom (at least from the time of the Tudors) to name the rooms of inns, these names often occurring in inn scenes in the works of contemporary dramatists; the custom is commented upon by Professor A. E. Richardson in his book *The Old Inns of England* (p. 74). In 'Mrs. Stanuell Day Accompt' (fol. 8) relating to the 'Red Lion Inn' is the following: '1800. April 5 to 10 Setting Range in Unicorn', and in Mr. Harrison's account for the same inn (fol. 138): '1806, 2 to 7 Feb, plastering Dado in Lyon'; these are apparently the first references relating to the naming of inn rooms in Doncaster.

Again in Mrs. Stanuell's account for the Red Lion there are several references to the 'Soldiers Room', 'March 2 1799 Mending floor in Soldiers Room' and 'Setting oven in Soldiers Room'. It is suggested that this term is not a romantic survival, but must be taken literally. At this period soldiers were not regarded as fit subjects for the ordinary person to mix with, especially in inns of a better class.

An intimate item is recorded in 'Mr. Moore Surgeon Day Accompt' (fol. 285). Among details of white-washing, Repairing Buildings, &c, there is an entry for the year 1810, but with no other dating; it reads 'Mr. Moore's bill for My Son Charles . . . . . 19/-'. Could it be that the 19/- was the payment (belated perhaps) marking 'Son Charles' advent into 'this troublous life'. Mr. John Moore, surgeon, was a son of the Revd. Stephen Moore, Vicar of Doncaster (1790-1807). John married a daughter of Frances Withers and Oswald Allen, surgeon, of York; Frances Withers was a collateral descendant of the Revd. Edmund Withers, master of Doncaster Grammar School (1707-37).[63]

The account book also illustrates another aspect of the domestic economy of the period, namely the keeping of pigs by many of the townsfolk, from the gentry down to the humble artisan. There are frequent references to the building of 'pig-coats' and even Miss Murphy's ladies' academy had its pig-fold (fol. 229). A few among the more affluent kept their own cows, often in quarters which today would be regarded as uncomfortably close to their houses. The keeping of cows entailed the provision of a dairy, a circumstance to which we have previously referred (see Mr. Wroughton's account).

Some of the town houses also possessed a dove-cote, its inmates providing a cheap and tasty addition to the dining-table, especially as the birds would be fed principally at the expense of neighbouring farmers. On folio 45 we find Thomas III's bill for providing the Revd. John Ramsden with one of these useful adjuncts:—

Decr 17 to 22.	making Dove Coat 2 Do Varley	0. 5. 0
	2 Do Lumley 1 Barrw Morter	0. 4. 6.
	3 Barrw Lime hair 70 bricks	0. 4. 8¼
	4 Metts plaster 2 B of Reeds	0. 3.10
	To 100 of 6d Nailes	6d.

Before the setting up of a police force, night watchmen were supposed to keep vigil, boxes being provided for them at various locations in the town. In the Corporation accounts there is an

30 (left) *The Staircase at Southwold House, Warter. The house was considerably altered and improved by William Peter Anelay for the Marquess of Normanby in 1974. Architect, Francis F. Johnson.*

31 (right) *The Octagon Tower, Studley, and Fountains Estate in course of rebuilding and restoration for the County of North Yorkshire, by William Peter Anelay in 1974. Architect-in-charge, W. T. C. Walker.*

32 (above) *Some Doncaster members of the firm taken in 1897. Walter I stands far left, and at the far right, in bowler hat and apron, is the foreman joiner, Frank Clegg. Standing next to him is Jack Chantrey, a scaffolder who worked for Anelays for his full working life.*

33 (below) *Members of the firm in 1972. Seated left to right, Mr. C. F. Oldfield (Joiners' shop Foreman), Mrs. J. Longbottom (Clerk), Mr. H. Tyerman (Joiner Foreman), Mrs. S. Hargreaves (Secretary), Mr. & Mrs. William Anelay II, Mr. & Mrs. William Peter Anelay, Mr. P. M. Johnston (Contracts Director), Mr. K. G. Greenley (Contracts Supervisor), Mr. F. Staples (General Foreman), Mr. L. Letby (Stonemason Foreman). Sitting on the ground are Helen, Charles and Elizabeth Anelay.*

entry dated 21st March, 1812 'By do pd Geo: Harrap for 6 New Watch Boxes Fixing &c as p, ... 27. 15. 0'.[64] The 'Corporation Day Acct for Rossington' (fol. 264) in the Anelay account book, contains an item which obviously referrs to Doncaster: '1816, August 23 to 31. . . . . . Removing Watch Box at the Waterhouse . . . . .'. Where the other five watch boxes were situated cannot be said, but from the above it is apparent that there was one at the Waterhouse, presumably placed there so that the Watchman, in his waking moments, could espy any suspicious stranger entering the town, either from the London or Thorne roads.

In addition to its intersecting roads, Doncaster was served by another highway, the River Don, and its associated canals; this means of relatively cheap transport was especially important to the builder, who relied upon it for the carriage of much of the lime and stone essential to his business. Reference has been made previously, in the account for the 'Chapell' house, to the payment of freight and lock dues; a further instance of the use of water transport occurs in 'Mr. James Jackson's Day Accompt' where stone is being brought to build a wall, 'April 4 1799 Paid the Keel-man for freight . . . . . 0. 9. 0., for Ale . . . . . 0. 2. 0.'

This brief survey of the social scene is ended as it began, with the master builder, now joined by his sons Thomas IV, and Charles. Towards the end of his book, Thomas III records their daily labour in a set formula, 'My Son, and Charles', and there are but few instances where this pattern varies. One of these will be found in the account of Messrs. Woodhead and Hurst, for the Alterations to the Mansion House in the year 1831 (fol. 311) where his two sons, then out of their apprenticeship, are receiving the full craftsman's rate of 3/6 a day, while their father then approaching the age of 70, is content to take the humbler part and enter 'Self as Labourer 7 Days at 2/4.'

Doncaster—A Colourful Town

The Exterior and Interior Decoration of the Buildings of Doncaster as depicted in the Account Books of Messrs. Anelay

The principal material used in building decoration at the close of the 18th century was a coloured lime-wash with which a glue size was mixed. Glue is frequently mentioned in the accounts along with the pigments, and cost 1/- a pound. It is apparent that white-washing and colouring was carried out by the builder's men at this time, many examples being recorded.

Probably the majority of Doncaster's buildings were colour-washed on the outside, and particularly those which belonged to the Corporation. This colouring seems to have been renewed at fairly frequent intervals, the Grand Stand and other race-course buildings receiving special attention in September, before the Race-meeting. Although it is not absolutely certain, it would seem that the prevailing exterior colour was ochre, which cost 4d lb in 1801 (fol. 149), falling to 3d lb in the year 1825, when 5 lbs were charged in the Parish of Doncaster Day Accompt for the colour-washing of the Work-house (fol. 296).

From the earlier accounts, 1798-1802, it is possible to obtain some idea of the colour schemes employed. In Mr. Firth's house where Thomas III had earlier been 'Casting and putting up Ornaments in Blocks', the staircase and bedrooms had been coloured using '2 Stone paris white at 1s', '10 lbs of pink at 8d' and '7 lbs of Oker at 4d', the front of the house and the stable being coated with a wash of pale blue for which '2 Stone paris white, prussia blue' were used at a cost of 4/3. No doubt the proportions of the mixture altered, but in a further entry in the second Day

Book dated 31st May, 1800, the quantities are stated to be '6 ounces prusia blue to 1 stone parris white', the 'prusia blue' costing 9d an ounce. At this price it is evident that a little had to go a long way.

In Mr. Lindley's Accompt for materials there is a list of colours employed in the re-decoration of the Mansion House which is unfortunately incomplete, many of the prices for colours being omitted. The list details Paris White 10½ stones, 4 ounces of Prusia blue, ½ ounce of indigo, 6 pounds of copperis [copperas], 2½ pounds of Dutch Pink, and a ¼ pound of Lamp Black, in addition to these, ½ pound wax and a similar quantity of Rosin, along with 13 lbs of glue is specified. How these colours were mixed does not appear, but it would seem that in some instances oil was used, as a quantity is charged at 1/3, but what this monetary figure may represent in terms of quantity is uncertain. Also in the above account is an item for spetches, of which 30 lbs were charged. These were scrap leather ends which were boiled down to make glue.

In other accounts in the book there are more details relating to the prices of pigments used. In Mr. Wroughton's Day Accompt dated January 1802 there are a series of entries under the heading, 'Colouring at the House lately Occupied by Dr. Stewart'. The work occupied a bricklayer and his labourer nine days, in addition to which the master was engaged on the work for two days. The pigments used included the inevitable 'parris white' of which 3 stones were charged at 1/-, 9 lbs of Dutch Pink at 8d lb, 2 lbs of Green at 1/4 lb, 2 ounces of indigo at 6d, 4 lb of Lamp Black at 4d lb, and 5 lbs of glue at 1/- lb, the total charges for the work amounting to £3.12.1. The Doctor Stewart referred to was Thomas Ruddiman Stuart, who had been on the staff of the Sheffield General Infirmary, and who came to live in Doncaster. The house, built by William Lindley, is still standing on the south corner of Princes Street at its junction with Hallgate. According to contemporary books of *Builders' Prices*, this type of work was usually undertaken by the plasterer, but Thomas III almost without exception seems to have employed bricklayers, as previously stated.

Decorating or colouring work was often carried out by measure. This was the case when the Town Hall was re-decorated in 1800, although the items occur in the Day Account, and are as follows:—

1800 April 4th to 19th	To 270 yds of White-washing at 1d per yd.	1.2.6.
	To 30 yds of Cornice white-washing at 1d.	2.6.
	To 109 yds of Green wash at 5d	2.5.5.
	To 517 yds of Yellow wash at 2d per yd.	4.6.2.

Although it is nowhere explicitly stated, it is possible that the difference in price between the white and the green or yellow wash may be due to the fact that the cost of the materials is included in the price as well as the labour costs.

One of the latest Corporation accounts in the book is still recording in the year 1841 a general smartening up of Corporation property immediately before the opening of the Race Meeting, from 2nd August to 1st September. Thomas III's men were busy white-washing and colouring the Mansion House kitchens and store rooms, 'the Spectators' Room at the Stand', the outside of the Town Hall, the Stewards' Stand; the posts and rails on the race-course were white-washed, while the old offender the Water-house was 'Scraped and Repaired' before being coloured.

COMPARISON OF DAILY WAGES AND OF BUILDING MATERIALS

Quakers' Meeting House (1798), folio 2, *et seq.*

Folio	DESCRIPTION	ANELAY ACCOUNT BOOK	BUILDER'S PRICE BOOK
2	Bricklayer Rate per Day	2/6	4/– 4/6*
2	Labourer do	2/–	2/9 3/–*
2	Ceiling Laths	1/2 per bunch	2/3 per bundle
2	Pantiles	¾ ea. 6/3 per 100	13/6 per 100
3	Chimney pots	3/–	6/6

Sir Geo. Cooke's House (1801) 2nd fly-leaf

Folio	DESCRIPTION	ANELAY ACCOUNT BOOK	BUILDER'S PRICE BOOK
2nd fly-leaf	Bricklayer Rate per Day	2/6	4/– 4/6*
,,	Labourer do	1/10	2/9 3/–*
,,	Pan Tiles	8/4 per 100	13/6 per 100
,,	Hair (per mett of 2 bushels)	2/3	5/6

Mr. Rawstorne's Day Account (1798-1799)

Folio	DESCRIPTION	ANELAY ACCOUNT BOOK	BUILDER'S PRICE BOOK
43	Bricklayer Rate per Day	2/6	4/– 4/6*
43	Labourer do	2/–	2/9 3/–*
43	Mortar per barrow	6d	10½d
43	Lime-hair per barrow	10d	1/3
85	Laths	1/4 per bunch	2/3 per bundle
85	Hair (per mett of 2 bushels)	2/3	5/6

Chapell House (1803)

Folio	DESCRIPTION		ANELAY ACCOUNT BOOK	BUILDER'S PRICE BOOK	
10	Bricklayer	Rate per Day	3/6	4/–	4/6*
10	Labourer	do	2/9	2/9	3/–*
10	Tiles (1d each)		8/4 per 100	13/6 per 100	
245	Mortar	per barrow	7d	10½d	
245	Lime-hair	per barrow	1/–	1/3 per barrow	
246	Putty	per pail	6d	1/9	

* *Note:* The second figure is the rate from Lady Day to Lord Mayor's Day, which obviously applied to London. There is no indication whatever in the Anelay Account Books that a similar differential rate was paid by him. The measures given in Crosby's *Builders' New Price Book* for 1804 have been adjusted to the Anelay measures.

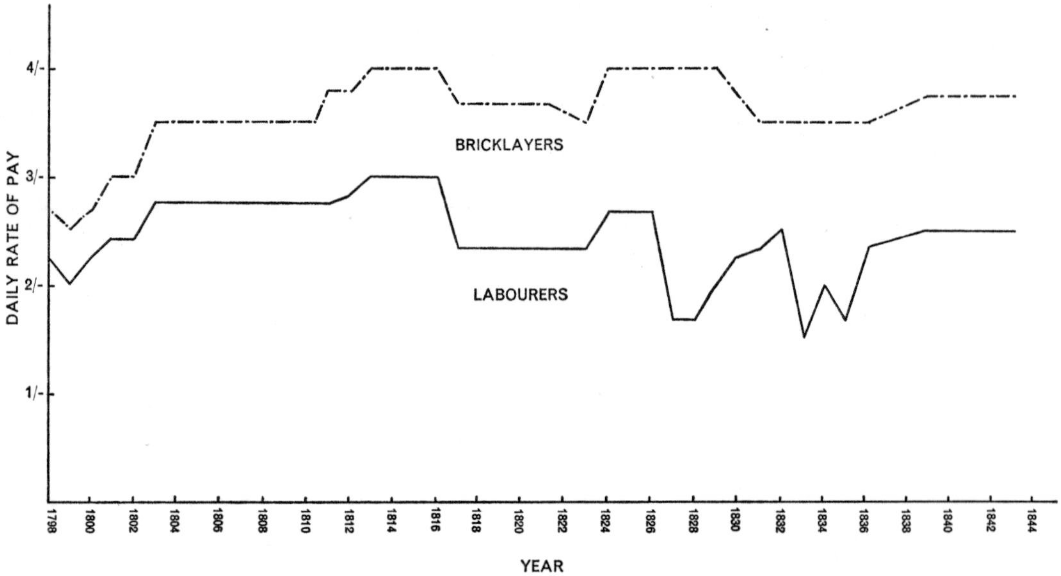

Rates of Pay, 1798-1844

NOTES

1 Cash Book 1772/3-1783, fol. 6. Reference Library, Doncaster Central Library
2 Cash Book 1772/3-1783, fol. 5. Reference Library, Doncaster Central Library
3 Calendar Doncaster Records, Vol. IV, p. 186
4 Cash Book 1772/3-1783, fol. 6. Reference Library, Doncaster Central Library
5 Cash Book 1772/3-1783, fol. 18. Reference Library, Doncaster Central Library
6 Wm. Sheardown, *Moat and Mound. A Defence of the Town of Doncaster* (1869), p. 32.
7 Cash Book 1772/3-1783, fol. 13. Reference Library, Doncaster Central Library
8 Cash Book 1772/3-1783, fol. 12. Reference Library, Doncaster Central Library
9 Cash Book 1810-1817, fol. 68. Reference Library, Doncaster Central Library
10 Cash Book 1783-1797, fol. 60. Reference Library, Doncaster Central Library
11 Calendar Doncaster Records, Vol. IV, p. 241
12 Edward Miller, *History and Antiquities of Doncaster*, 1804, p. 184
13 Credit Book 1772/3-1782, fol. 107. Reference Library, Doncaster Central Library
14 C. W. Hatfield, *Historical Notices of Doncaster* Vol. II, p. 367
15 *Doncaster, Nottingham and Lincoln Gazette*, 19th December 1806
16 Calendar Doncaster Records, Vol. IV, p. 265
17 Cash Book 1783-1787, fol. 41. Reference Library, Doncaster Central Library
18 Ledger 1797-1808, fol. 110. Reference Library, Doncaster Central Library
19 Calendar Doncaster Records, Vol. IV, p. 265
20 Calendar Doncaster Records, Vol. IV, p. 265
21 Cash Book 1783-1797, fol. 36 *et seq*. Reference Library, Doncaster Central Library
22 Cash Book 1783-1797, fol. 61 *et seq*. Reference Library, Doncaster Central Library
23 Calendar Doncaster Records, Vol. IV, p. 265 *et seq*
24 Credit Book 1772/3-1783, fol. 95. Reference Library, Doncaster Central Library
25 Engraving, York Public Library
26 Ledger 1797-1808, fol. 131. Reference Library, Doncaster Central Library
27 Ledger 1797-1808, fol. 154. Reference Library, Doncaster Central Library
28 Calendar Doncaster Records, Vol. IV, pp. 268-9
29 Calendar Doncaster Records, Vol. IV, p. 269
30 Ledger 1797-1808, fol. 131. Reference Library, Doncaster Central Library
31 C. W. Hatfield, *Historical Notices of Doncaster*, Vol. II, p. 153
32 *Peeps into the past, being passages from the diary of Thomas Asline Ward*, p. 97
33 Ledger 1797-1808, fol. 144. Reference Library, Doncaster Central Library
34 Calendar Doncaster Records, Vol. IV, p. 270
35 Calendar Doncaster Records, Vol. IV, p. 271
36 *Doncaster, Nottingham and Lincoln Gazette*, 5th September 1806
37 Ledger 1797-1808, fol. 127. Reference Library, Doncaster Central Library
38 Ledger 1797-1808, fol. 24. Reference Library, Doncaster Central Library
39 Ledger 1797-1808, fol. 80. Reference Library, Doncaster Central Library
40 Ledger 1797-1808, fol. 127. Reference Library, Doncaster Central Library
41 Calendar Doncaster Records, Vol. IV, p. 280
42 Calendar Doncaster Records, Vol. IV, p. 280
43 Cash Book 1810-1817, fol. 97 *et seq*
44 *Note:* The pillars of this portico were actually 'Tuscan', but they are described as 'Doric' in the newspaper account
45 Calendar Doncaster Records, Vol. IV, p. 260
46 Wm. White, Directory West Riding, Vol. I, 1837
47 Wm. Sheardown, *Doncaster Races, Past and Present*, p. 13
48 *Doncaster, Nottingham and Lincoln Gazette*, 24th October 1828
49 *Doncaster, Nottingham and Lincoln Gazette*, 20th December 1833
50 J. Tomlinson, *Doncaster from the Roman Occupation to the Present Time*, p. 226
51 This refers to the Gaol of 1828-29, designed by Woodhead and Hurst, and built by Messrs. Lockwood, which was also grossly under-estimated
52 E. Miller, *History and Antiquities of Doncaster*, annotated by Wm. Hurst, ii verso between pp. 170 and 171. Leeds Reference Library
53 J. Tomlinson, *Doncaster from the Roman Occupation to the Present Time*, p. 265
54 Cash Book 1783-1797, fol. 85. Reference Library, Doncaster Central Library
55 *Complete works of Benjamin Thompson, Count von Rumford*, 4 vols. Boston 1870-75, and Dictionary of National Biography, Vol. LVI, p. 205
56 Baines, Directory of West Riding 1822, p. 176. John Pollard, plasterer, Duke Street
57 Yorkshire Marriage Registers, West Riding, Vol. IV, p. 272. Ed: E. K. Blumhardt
58 Wm. White. West Riding Directory, Vol. II, p. 259
59 Sykes MS., XIX. Reference Library, Doncaster Central Library
60 Baines, Directory West Riding, p. 176
61 Cash Book 1783-1797, fol. 141. Reference Library, Doncaster Central Library
62 J. E. Jackson. *History of St. George's Church, Doncaster*. App. VIII, p. 6
63 *Doncaster, Nottingham and Lincoln Gazette*, 22nd August 1851
64 Cash Book 1810-1817, fol. 34. Reference Library, Doncaster Central Library

PART FOUR: LIST OF ANELAY BUILDINGS

THERE ARE many gaps in the list of the Anelay work due to lack of records of the earlier years in Doncaster. The pre-1914 papers went for salvage in the First World War. In preparing this list a great debt is owed to Mr. James E. Day of Doncaster, the author of Part Three of this book; to Mr. Cecil Leckenby of York who abstracted from his firm's Contract Register the works which Anelays carried out for Walter H. Brierley, 1900-18, and to Mr. William Peter Anelay who combed his firm's more recent records for works done between 1919 and 1974.

Walter H. Brierley had many special techniques in the way he built and employed John Vaux as his special Clerk of Works to make sure that this high degree of craftsmanship was carried out. When Anelays were the contractors, John Vaux delighted in recounting how superbly they always rose to the occasion. He also repeated what may be a slightly apocryphal story of how the basic techniques used by the Ancient Monuments Department of the then Ministry of Works came to be synthesized under Sir Charles Peers. The story is that Charles Peers, as he then was, observed the Brierley/Anelay restoration techniques being employed at Middleham Castle, Clifford's Tower and Bolton Castle, and found them so excellent that he adopted and codified them to be the standard practice for the Department. From a visual inspection it seems that there may be some substance in this story.

JOHN THOMPSON

c. 1745	Doncaster: The Mansion House. The nature of the work done is undefined. The main building periods of the Mansion House are 1745-48 with James Paine (c. 1716-89) as architect, and 1801 and 1806 with William Lindley (c. 1739-1818) as architect.
1748	Doncaster: Laying Corporation pavement and maintenance until 1755.
1753	Doncaster: Constructing a causeway between Grey Friars and Mill Bridge in association with Henry Walker.

THOMAS ANELAY II

1772-88	Rossington: West Riding, Yorkshire. 'Much of the village built' and large quantities of bricks made and fired in kilns built and repaired by Anelays and operated by Doncaster Corporation.
1773	Doncaster: Erection of ornamental wall in the Corn Market.
1773	Doncaster: Arching of the Bardike Sewer.
1778	Doncaster: A small undefined work at the Mansion House.
1779-80	Doncaster: Building the new Gaol in St. Sepulchre Gate. Architect, William Lindley.
1778	Rossington: West Riding, Yorkshire. The new Court House built.
1793	Doncaster: Building the Dispensary in association with George Hewitt. Architect, William Lindley.
1798	Doncaster: Friends' Meeting House, formed out of a granary.

THOMAS ANELAY III

1801	Doncaster: The Mansion House. Construction of the attic storey. Architect, William Lindley, c. 1739-1818. Lindley was in John Carr's office in York, and Peter Atkinson I is mentioned in Lindley's will 'as my late respected friend'. Towards the end of his life

Lindley was in partnership with Woodhead. William Hurst (1787-1844) was articled to this firm. The firm became Woodhead & Hurst of Doncaster, and in due course was joined by William Lambie Moffatt. The practice seems to have ceased about 1850.

1804 Doncaster: Racecourse. Alterations to Grandstand. Architect, William Lindley. Original architect, John Carr.

1805 Doncaster: Racecourse. Judges' and Stewards' Stand erected. Architect, probably William Lindley.

1808 Doncaster: School of Industry erected.

1814 Doncaster: Alterations to the Theatre. Architect, William Lindley.

1823 Doncaster: Erection of 42 Booths for Publicans at the Racecourse. Architect, William Hurst.

1828 Doncaster: Alterations to the Town Hall.

1843 Doncaster: St. George's Church. Completion of restoration of East end. This work was to the medieval church which was destroyed by fire in 1853.

THOMAS IV AND CHARLES ANELAY

1846 Hatfield: West Riding, Yorkshire. St. Lawrence's Church. Stonework restoration to this fine medieval church.

1847 Doncaster: The Market Hall. Architect, John Butterfield, the Corporation Steward. The new building was on the site of the Old Town Hall, which incorporated considerable remains of the 11th century Church of St. Mary Magdalene. There was much fuss about the demolition. The Corporation decided to proceed and when the demolition was finished 'The Architect then took possession of the ground and by the year 1849 the Town rejoiced in its new Market House'. J. E. Jackson, *History of the Ruined Church of St. Mary Magdalene*, Doncaster, 1853.

1849 Kirk Bramwith: West Riding, Yorkshire. Restoration of medieval church. Architect, probably the Rector, the Revd. W. H. B. Daubeny.

1849 Barnburgh Church: West Riding, Yorkshire. New stonework to two windows in chancel.

1850 Barnburgh Rectory: Arch itect, Thomas A. Hine of Nottingham. This highly romantic house has recently been de molished.

1851 Doncaster: Erection of Schoolroom for Yorkshire Institution for the Deaf and Dumb. Architect, William Lambic Moffatt.

1851-2 Doncaster: Christ Church Day Schools built. Architect, William Lambic Moffatt.

1853 Doncaster: Racecourse. The Gentlemen's, Trainers' and Reporters' Stand built.

1855 Doncaster: Cemetery Chapels, Lodge and Gateway. Architect, Robert James Johnson, 182?-92.

1856 Doncaster: Christ Church. New vestry and alterations to chancel. Architect, Sir George Gilbert Scott. Original Architect, William Hurst.

1858 Doncaster: Conversion of Militia Depot into West Riding Constabulary offices, Lock-up and Superintendent's residence.

1860 Doncaster: Improvements and alterations to the Vicarage for the Revd. Dr. Charles Vaughan.

1861 Doncaster: Alterations to the Grandstand. Architect, John Butterfield.

1861 Badsworth: West Riding, Yorkshire. Building a new wing to Badsworth Hall in association with Mr. Athron of Doncaster for R. H. Jones, Esq. Architect unknown.

1863 Doncaster: Covered Wool Market erected.

1864 Kirk Sandal: West Riding, Yorkshire. Restoration of St. Oswald's Church. Architect, J. M. Teale of Doncaster.

1865 Doncaster: The Infirmary, Wood Street. Architects, Brundell & Arnold.

1866 Doncaster: Gas works for the Great Northern Railway. Gas Engineers, Messrs. Knapton.

1867 Doncaster: Alterations and improvements at the Racecourse.

1868 Doncaster: Re-facing Beckett's Bank with stone. Architect, Benjamin Shaw Brundell, c. 1826-97.

1868	Doncaster: Building a stone-gabled house in Regent Square.
1868-70	Arksey Church, near Doncaster: Restorations of medieval church under the direction of Sir George Gilbert Scott.
1870	Doncaster: Church of St. John, Catherine Street, built as Chapel of Ease to Christ Church. Architect, Thomas Penrice of Clapham.
1873	Frickley Church: West Riding, Yorkshire. Restoration of medieval church under the direction of architect G. Williams, London.
1874	Doncaster: Carving a memorial tablet to Charles Baker, Esq., at the Yorkshire Deaf and Dumb Institution.

WILLIAM ANELAY I

1877	Doncaster: West Riding, Yorkshire. Taking up and levelling gravestones in St. George's churchyard.
1879	Doncaster: Rebuilding the Robin Hood Inn, Marsh Gate.
1882	Rossington: West Riding, Yorkshire. Rossington Hall built for James Brown, Esq. Architect not known but Samuel Sanders Teulon has been suggested.
1882	Tickhill: West Riding, Yorkshire. Restoration of the tower of the medieval Church of St. Mary.
1882	Doncaster: Racecourse. Alterations and additions to the Ladies' and to the Noblemen's Stands.
1883	Hickleton: West Riding, Yorkshire. Restoration of medieval church for Lord Halifax. Architects, George Frederick Bodley and Thomas Garner.
1884	Doncaster: Nos. 5, 6, 7 and 8 Herschell Terrace erected, now Nos. 9, 11, 13 and 15 Thorne Road. Architect not known.
1884	Finningley: Nottinghamshire. Restoration of the medieval Church of the Holy Trinity. Architect, Charles Hodgson Fowler.
1885	Stainforth: West Riding, Yorkshire. Cemetery chapel and lodge built. Architect, E. Dolby of Fulham.
1885-6	Wollaton: Nottinghamshire. Restoration of St. Leonard's Church and South Aisle added. Architect, Charles Hodgson Fowler.
1889	Doncaster: Free Library and Art School. Architect, H. A. Cheers of Twickenham.
c. 1890	Bawtry: West Riding, Yorkshire. Restoration of Bawtry Hall. Architects, De Maine & Brierley.
1891	Marr: West Riding, Yorkshire. Restoration of the medieval Church of St. Helen. Architect, Charles Hodgson Fowler.
1891	Doncaster: Building a private chapel at Hesley Hall for B. I. Whitaker, Esq. Architect, Henry Christian of London.
1894	Doncaster: St. Jude's Church, Hexthorpe. Architect, Lord Grimthorpe, 1816-1905, and Herbert Athron. . . . 'tracery in gawky Geometrical Style' (Pevsner). This is Grimthorpe's last complete building.
1894	Cantley Church: West Riding, Yorkshire. Extensive restorations and much fine furniture and fittings furnished with gilding and polychromatic decorations. Architect, Bucknall & Comper. The contrast between the coarse, dull vulgarity of St. Jude's Hexthorpe and Cantley's atmosphere, purity and vibrant colour would have been almost unbelievable.
c. 1895	Tickhill: West Riding, Yorkshire. 'Weardale' built. Architect not known.
1898	Doncaster: York City and County Bank at the junction of High Street and Baxtergate. Architects, James Demaine & Walter H. Brierley of York. James Demaine, 1842-1911 became a partner of the firm Atkinson & Demaine of York in 1877. This firm may be traced backwards through various partnerships to John Carr, 1723-1807, and forward through Walter H. Brierley, 1862-1926; James H. Rutherford, 1874-1946; John Stuart Syme, 1872-1958. Cecil Leckenby, David R. Leckenby, John Keighley and Keith Groom are the present partners. Anelays carried out and continue to carry out, much work for this immensely long-established practice.

THOMAS ANELAY VI

1900 York: Dringhouses. Erection of the Fox Inn. Architect, Samuel Needham (1868-1927), York.

For the years 1900-18 the Contract Register of Brierley & Rutherford appears to be the most detailed extant consecutive record of the Anelay works. The following works were executed during the Demaine, Brierley, Rutherford reign in the practice briefly described above.

1900 York: A new wing at the King's Manor (then the School for the Blind) now part of the University of York.

1900 York: Lodge at Grimston for J. J. Hunt.

1900 Castleford: West Riding, Yorkshire. St. Paul's Church, Glass Houghton.

c. 1900 Doncaster: West Riding, Yorkshire. Additions and alterations to Edenthorpe Hall, previously known as Street-thorpe Hall.

1901 York: Grimston Court, for J. J. Hunt.

1901 Holme Church: Yorkshire.

1901 Sprotborough: West Riding, Yorkshire. Farmhouse.

1902 York: St. John's College. New wing.

1902 Thorpe Underwood: West Riding, Yorkshire. The Hall for F. W. Slingsby—'in tasteful Tudor' (Pevsner).

1902 Bramham Church: West Riding, Yorkshire. Repairs.

1902 Hutton Buscel Church: North Riding, Yorkshire. Repairs.

1903 York: Fulford. House for C. A. Cooper.

1903 York: Clifford's Tower. Restoration.

1904 York: Fulford Church. Repairs.

1904 Gilling Castle: North Riding, Yorkshire. Alterations and repairs for W. Slingsby Hunter.

1904 Finghall Church.

1905 Middleham Castle: North Riding, Yorkshire. Restoration for Lord Masham.

1905 Whixley Hall: West Riding, Yorkshire. Alterations for Alfred Taylor.

1905 York: 'Bishopsbarns', St. George's Place. This is Walter Brierley's own house.

1906 Cantley Church: West Riding, Yorkshire. Repairs and refurnishing after fire (see work done 1894). The 1906 work is what the visitor now sees. Cantley has the first of Comper's 'English Altars'. The interior of this little church is one of the best of Comper's 'Unity by Exclusion'. Sir Ninian Comper, 1864-1960, was a major English ecclesiastical architect.

1907 Doncaster: Extension to West Paine Wing erected at Cusworth Hall. Architect not known.

1907 Rawcliffe: West Riding, Yorkshire. St. James Church. New Chancel. The body of the church 1841. Architects, Hurst & Moffat.

1908 Normanton Church: West Riding, Yorkshire. Repairs and additions.

1908 Middleham Castle: North Riding, Yorkshire. Restoration for Lord Masham.

1909-12 Gilling Castle: North Riding, Yorkshire. Repairs and alterations for W. S. Hunter.

1909 Low Catton Church: West Riding, Yorkshire. Restoration.

1909 Pickering: North Riding, Yorkshire. Beckett's Bank.

1909 York: Racecourse. Alterations.

1910 Pickering: North Riding, Yorkshire. House at Keldy Castle for P. B. Reckett. Architect, John Bilson, 1856-1943.

1910 Hackness: North Riding, Yorkshire. The Grange for K. F. Johnstone.

1910 Hackness: North Riding, Yorkshire. Rebuilding the Hall after fire. This house originally designed by Peter Atkinson I, 1796.

1910 Acklam Hall: North Riding, Yorkshire. For W. Hustler-Hustler.

1910 York: St. Luke's Church. Architect, Walter Brierley. Remains unfinished. In design compares with Rawcliffe chancel (also detail).

1911 Helmsley: North Riding, Yorkshire. Beckett's Bank.

1911 Foston Church: North Riding, Yorkshire. Restoration and additions.

1912 Sledmere House: East Riding, Yorkshire. Minor works.

1912 Arncliffe Hall: North Riding, Yorkshire. Rebuilding after fire. This is a John Carr House, 1753-4.

1912 Saltmarshe Hall: East Riding, Yorkshire. Alterations, built c. 1820.

1913 South Kirkby Church: West Riding, Yorkshire. Restoration.

1913 Tadcaster: West Riding, Yorkshire. Oxton Hall. Rebuilding after fire.

1913 Malton: North Riding, Yorkshire. Strong-room at Beckett's Bank.

1914 Skelton Castle: North Riding, Yorkshire. Alterations and additions. Possibly there are some medieval bits but mostly a castellated effort of c. 1794. Some work by Sir John Soane.

1914 Hackness Hall: North Riding, Yorkshire. Reconstruction of kitchen wing.

1914 Tockwith: West Riding, Yorkshire. Alterations to Vicarage.

1914 Doncaster: Alterations and additions to Beckett's Bank.

1914 York: Alterations to The Yorkshire Club.

1915 Kellington: West Riding, Yorkshire. Alterations to Vicarage.

1915 Malton: North Riding, Yorkshire. Alterations to London Joint Stock Bank.

1916 York: Assize Courts. Drainage.

1916 York: Workshops and Factory, Bishophill, for T. Cooke & Sons, Limited. Nice Arts and Crafts design.

1916 Jervaulx Abbey: North Riding, Yorkshire. New wing to the house for W. L. Christie, Esq.

1916 Norton: East Riding, Yorkshire. Wall, piers and gates to St. Peter's Church, designed by Charles Hodgson Fowler.

1916 Gilling: North Riding, Yorkshire. Farmhouse and cottages for W. G. Hunter.

1918 Harrogate: West Riding, Yorkshire. Alterations and additions to Park Place.

1918 Goathland: North Riding, Yorkshire. Alterations to the Orchards.

1919 York: Alterations and renewals at The Racecourse. Architects, Brierley & Rutherford.

1919 Castle Bolton: North Riding, Yorkshire. Bolton Castle, repairs. Architects, Brierley & Rutherford.

1919 Scarborough: Alterations to 'Graylands' for Thomas Hall. Architects, Brierley & Rutherford.

1920 York: Annexe to St. John's College. Architects, Brierley & Rutherford.

1920 Mirfield: West Riding, Yorkshire. New wing to the College of the Community of the Resurrection. Architect, John Bilson.

1921 York: Alterations to Middlethorpe Manor. Architects, Brierley & Rutherford.

1921 Nunnington: North Riding, Yorkshire. Additions and alterations to the Hall. Architects, Brierley & Rutherford.

1921 Leyburn: North Riding, Yorkshire. Sorrelsykes Park. Alterations and additions. Architects, Brierley & Rutherford.

1922 Knaresborough: West Riding, Yorkshire. Alterations and repairs to Goldsborough Hall for Viscount Lascelles. Architect, Sidney O. Kitson, 1871-1937.

1923 York: Restoration of the Merchant Adventurers' Hall. Architects, Brierley & Rutherford.

1923 Escrick: East Riding, Yorkshire. Restoration of the church after fire. Architect, John Bilson. Church built 1856-7; Architect, Francis Cranmer Penrose, 1817-1903.

1924 Malton: North Riding, Yorkshire. Alterations and additions to Easthorpe Hall for Lord Grimthorpe. Architects, Brierley & Rutherford. This fine house is attributed to John Carr.

1924 York: Repairs to the late medieval Bedern Chapel. Architects, Brierley & Rutherford.

1925 Rotherham: West Riding, Yorkshire. Firbeck Hall. Reinstatement after fire. Architects, Brierley & Rutherford.

1926 York: 'Goddards', Tadcaster Road, for Noel G. Terry. Architects, Brierley & Rutherford. Compare with Brierley's own house (1905) and The Garth, Bishopthorpe.

1926 Richmond: North Riding, Yorkshire. Holy Trinity Church. Repairs to the tower. Architects, Brierley & Rutherford.

1926 York: Coney Street. Building branch for Liverpool and Martins Bank. Architects, Brierley & Rutherford.

WILLIAM ANELAY II

1927	Ripon: West Riding, Yorkshire. Alterations and repairs to Fountains Hall for Commander Clare Vyner and repairs to the ruins of Fountains Abbey. The House, built c. 1611, with materials from Fountains Abbey. In design it may be considered in the Smythson's Derbyshire manner.
1928	York: Alterations to Deighton Grove for C. N. De Yarburgh Bateson. Architects, Bland & Bown.
1929	Richmond: North Riding, Yorkshire. Roman Catholic Chapel at Catterick Camp. Architects, Leonard Stokes & Drysdale.
1929	Filey: East Riding, Yorkshire. Repairs to stonework of Parish Church. Architect, John Bilson.
1930	Harewood House: West Riding, Yorkshire. Alterations for the Earl of Harewood. Architects, Brierley & Rutherford. The House, 1759, by John Carr with interiors by Robert Adam and much re-casting by Sir Charles Barry.
1931	Redcar: North Riding, Yorkshire. Church of All Saints, Dormanstown. Architect, Leslie T. Moore, 1883-1957. Very competent design for very small cost and comparable with the church, St. Hilda's Shiregreen, Sheffield, which he designed.
1931	York: St. Martin le Grand, Coney Street. Stonework repairs. Architects, Ward & Leckenby.
1931	York: All Saints Church, North Street. Repairs to tower and spire. Architects, Brierley & Rutherford.
1931-6	Doncaster: St. George's Church. Designed by Sir Gilbert Scott, 1854-8. Architects for the repairs, Walker & Thompson of Doncaster.
1932	Castle Howard: North Riding, Yorkshire. 17th, 18th and 19th centuries. Architects, Vanburgh, Hawksmoor and Robinson. Restoration: Architects, Brierley & Rutherford.
1933	Doncaster: Church of St. James, designed by Lord Grimthorpe, 1858. Architects for the repairs, Walker & Thompson.
1933	Aldwark: North Riding, Yorkshire. St. Stephen's Church, designed by Edward Buckton Lamb, 1846. Architects for the repairs, Ward & Leckenby.
1934	West Heslerton: East Riding, Yorkshire. Alterations to The Hall for Captain Cuthbert Dawnay. Architects, Ward & Leckenby.
1934	York: Repairs to the old church of St. Oswald, Fulford. Architect, F. T. Penty.
1935	Elvington: York, East Riding, Yorkshire. Alterations and additions to Elvington Hall for Major J. R. Bury-Barry. Architect, Guy D. Channon.
1936	York: Alterations to Middlethorpe Manor for Sir Frank Terry. Architects, Ward & Leckenby. Very late 17th century with 18th century additions.
1936	Methley: West Riding, Yorkshire. Stonework and roof repairs to medieval church. Architect, Guy D. Channon.
1937	Thirsk: North Riding, Yorkshire. Alterations and additions to Kepwick Hall, 1873, with additions in 1889, for J. E. Guthe. Architects, Brierley, Rutherford & Syme.
1938	Harrogate: West Riding, Yorkshire. Reinstatement of Nidd Vicarage after fire. Architect, Guy D. Channon.
1938	Norton: East Riding, Yorkshire. Hyde Park Hotel for Charles Rose & Co. Ltd. Architect, Guy D. Channon.
1938 and 1947	Harrogate: West Riding, Yorkshire. Repairs to Christ Church. 1831 Architect, John Oates; 1861 Architects, Lockwood and Mawson. Architect for repairs, Guy D. Channon.
1939	Ripon: North Riding, Yorkshire. Masham Vicarage. Architect, Guy D. Channon.
1940-3-5	Castle Howard: North Riding, Yorkshire. 17th, 18th and 19th centuries. Architects, Vanburgh, Hawksmoor & Robinson. Restoration Architects, Ward & Leckenby.
1940-1	York: Underpinning and repairs to Assembly Rooms. 1730 design by Lord Burlington; portico 1828, Prickett & Watson. Architects for the repairs, Ward & Leckenby.
1940-45	York and East Riding, Yorkshire: Term contracts, etc., for the War Department.
1942	York: War damage repairs for York Corporation.
1945	York: 4 houses in Water Lane Estate for York Corporation. Designed by City Architect.

1945	Northallerton: North Riding, Yorkshire. Repairs to roofs of Ainderby Steeple Church (medieval). Architect, Guy D. Channon.
1945	Ripon: North Riding, Yorkshire. Repairs to roofs of Holy Trinity Church. 1826 Architect, Thomas Taylor. Architect for the repairs, Guy D. Channon.
1946	Sheriff Hutton: North Riding, Yorkshire. Repairs to Sheriff Hutton Park for Admiral J. Egerton. Architect, Guy D. Channon.
1946	York: St. Olave's Church. Stonework repairs for medieval 16th and 19th century church. Architect for repairs, G. C. Ferrey.
1946-7	Malton: North Riding, Yorkshire. 8 houses at Great Habton, 10 houses at Hovingham for Malton R.D.C. Architect, Guy D. Channon.
1947 and 1953	York: All Saints Church, Pavement. Repairs to roofs, stonework of tower, etc., of medieval church. Architects for the repairs, Brierley, Syme & Leckenby.
1948	Escrick: East Riding, Yorkshire. Alterations and additions to Escrick Hall to form Queen Margaret's School. Architects, Needham & Thorp. The Hall, c. 1680-90, much altered and having work by John Carr (?) and Edward Blore.
1948	Little Ouseburn: West Riding, Yorkshire. War damage repairs to the medieval church. Architect for the repairs, George G. Pace.
1948	Newton-on-Ouse: North Riding, Yorkshire. War damage repairs to spire and tower of the church, 1846. Architect, G. T. Andrews. Architect for the repairs, G. C. Ferrey.
1949	Selby: West Riding, Yorkshire. St. James' Church. New tower top and war damage repairs. Architects, Biscomb & Ferrey. Detailed drawing of pinnacles by A. Newton Thorpe.
1949	Malton: North Riding, Yorkshire. The Lodge. Repairs and alterations for Earl Fitzwilliam's Malton Estate. Architect, Guy D. Channon.
1949	Armagh Cathedral: Northern Ireland. Furnishings for the Royal Irish Fusiliers Chapel in the south transept. Architect, George G. Pace.
1950	York: Holy Trinity, Micklegate. Repairs to medieval roof. Architect, George G. Pace.
1950	Kippax: West Riding, Yorkshire. Stonework and roof repairs to medieval church. Architect, Guy D. Channon.
1950	Malton: North Riding, Yorkshire. Huttons Ambo Hall. Repairs and alterations. Architects, Fairweather & Jordan.
1950	Tadcaster: West Riding, Yorkshire. Colton Lodge. Alterations and repairs for A. Yates, Esq. Architects, Ward, Porteus & Ruddick.
1950-6	Goodmanham: East Riding, Yorkshire. Hall Garth. Repairs and alterations and additions for Lady de Grimston. 1823 Architect, Charles Mountain of Hull. Architect for the repairs, Francis F. Johnson.
1951	Thirsk: North Riding, Yorkshire. Repairs to 15th century nave roof of the Parish Church. Architects, Ferrey & Mennim.
1951	Wakefield: West Riding, Yorkshire. Chapel of St. Mary on the Bridge. Repairs to stonework and roof. Medieval and 1847, Sir Gilbert Scott; and 1939, Sir Charles Nicholson. Architect for the repairs, George G. Pace.
1951-3	Ripon: North Riding. Yorkshire, Diocesan Training College. New dining block and study blocks and alterations to existing building. Original building 1858 architect, J. B. Atkinson. Architects for the extension Ward, Ruddick & Ward.
1952-4	Darrington: West Riding, Yorkshire. Church. Repairs to medieval roof. Architect, George G. Pace.
1953	York: Judges' Lodging. Alterations and repairs for the Yorkshire County Committee. Architects, Kenneth Ward & Cecil G. F. Ward. Work continued to 1956 and further works done 1966-7 and in 1970 and 1972. A fine building of 1718-25.
1953	Eastrington: East Riding, Yorkshire. Repairs to medieval roof of church. Architect, George G. Pace.
1953	New Earswick: York. 8 flats for Old People. Architects, Needham, Thorp & White.
1953	Tadcaster: West Riding, Yorkshire. Stonework repairs to medieval tower of the Parish Church. Architects, Biscomb, Ferrey & Whipp.

1954	Bridlington: East Riding, Yorkshire. Restoration of Bayle Gate. Architect, Francis F. Johnson. This is the medieval gateway to the Priory. Further restorations were done in 1960 and 1969.
1954	Coxwold Church: North Riding, Yorkshire. Restorations. Architects, Brierley, Leckenby & Keighley. Further repairs, alterations and new furnishings in 1965 and 1970.
1954 and 1968	Bedale: North Riding, Yorkshire. Repairs to tower and roofs of medieval church. Architects, Charlton & Crowther.
1955	Sherburn in Elmet: West Riding, Yorkshire. Restoration of the medieval roofs of the Parish Church (also in 1960-1). Architect, George G. Pace.
1955, 1961 and 1967	Alne: North Riding, Yorkshire. Parish Church, repairs to roofs, tower, etc.; medieval and 18th century. Architects, Biscomb, Ferrey & Mennim.
1955	York: Dringhouses. Dual purpose Church/Hall of St. James the Deacon. Architect, George G. Pace.
1956	Birmingham: Fittings for the War Memorial Chapel in the Parish Church of St. Martin-in-the-Bullring. Architect, George G. Pace.
1956	Burton Agnes: East Riding, Yorkshire. Stonework repairs to Burton Agnes Hall for M. Wickham Boynton. Architect, Francis F. Johnson. Further repairs in 1958 and 1961.
1956	Llandaff Cathedral: Wales. Fittings for the Welsh Regiment Chapel. Architect, George G. Pace.
1956	Fishlake: West Riding, Yorkshire. St. Cuthbert's Church. Repairs to medieval roofs. Architect, George G. Pace.
1956	Pocklington Grammar School: East Riding, Yorkshire. Extension to assembly hall. Architects, Ward, Ruddick & Ward.
1956	York: St. Martin-cum-Gregory. Roof and stonework repairs to medieval church. Architect, George G. Pace.
1957	Durham Cathedral: St. Gregory's Chapel. Furnishings. Architect, George G. Pace.
1957	Wortley: Leeds. St. John the Evangelist. New vestry and west window. Architects, Braithwaite & Jackman.
1957	Ebberton Hall: North Riding, Yorkshire. Restoration for W. De Wend Fenton. Architect, Francis F. Johnson. Built 1718, Architect, Colen Campbell.
1957	Ripon: West Riding, Yorkshire. Repairs and alterations to Newby Hall, stables and lodges for the New Hall Estate. Further works in 1963, 1968 and 1970. Architects, Ward, Ruddick & Ward. The house begun 1705; some interiors by Robert Adams who also designed the stables and lodges.
1958	Bramham Park: West Riding, Yorkshire. Restoration of Temples and Monuments for Bramham Park Estate. Architect, George G. Pace. Magnificent early 18th century grounds, park and exterior of the Mansion House and stables. Two architects have been suggested, Giacomo Laoni and Thomas Archer: much work by Detmar Blow, 1907.
1958	York: County Assize Court. Repairs and alterations for Yorkshire County Committee. Architects, Kenneth Ward and Cecil G. F. Ward. Fine 18th century building by John Carr with additions by P. Atkinson II, c. 1812.
1958	Darfield: nr. Barnsley, West Riding, Yorkshire. Stonework repairs to medieval church. Architect, George G. Pace.
1958	York: Dorothy Wilson's Hospital, Foss Bridge, 1812. Restoration and alterations for the trustees. Architects for the repairs, Ward, Ruddick & Ward.
1958	Helmsley: North Riding, Yorkshire. Nawton Tower. Additions and alterations for the Earl of Feversham. Architects, Needham, Thorp & White.
1959	York: Ingram's Hospital. Alterations and repairs for the Ings Property Company. Architects, Needham, Thorp & White. Built c. 1635-40 and incorporating a Transitional doorway said to have come from Holy Trinity Priory, Micklegate.
1959	Thorpe Underwood: West Riding, Yorkshire. Alterations to the Hall for F. E. Rhodes. Architects, Needham, Thorp & White.
1959	Beningbrough Hall: West Riding, Yorkshire. Repairs for the National Trust. Architects, Colin Rowntree & Partners. Additional work in 1965. Architects, Hunter, Stoker &

Mellor. The design of the Hall is attributed to Vanburgh, but evidence points to William Thornton, early 18th century.

1959 Bugthorpe: East Riding, Yorkshire. Dry-rot eradication in medieval church. Architect, Guy D. Channon.

1959 Scarborough: North Riding, Yorkshire. St. Mary's Church. East window and chancel paving. Architect, George G. Pace.

1960 Ripon: West Riding, Yorkshire. Restoration of Markenfield Hall for Lord Grantley. Architect, R. B. Armistead. A fine, partly-moated and fortified medieval house with later additions and two 17th century lodges.

1960 Ripon: West Riding, Yorkshire. Repairs to stonework of the Town Hall. Built in 1801 and designed by architect James Wyatt.

1960 Garrowby Hall: East Riding, Yorkshire. Alterations for the Earl of Halifax. Architects, Needham, Thorp & White. Built in 1892-3 and designed by W. E. Tower who was partner of and successor to C. E. Kempe, the stained glass artist.

1960 Naburn: Bell Hall, East Riding, Yorkshire. Repairs and alterations for J. H. Baines, Esq. Architects, Ward, Ruddick & Ward.

1960 Bedale: North Riding, Yorkshire. Burneston Church, roof and dry-rot repairs to medieval church. Architects, Ward, Ruddick & Ward.

1960 Old Malton: North Riding, Yorkshire. Stonework repairs to Transitional Norman and Early English west front of church. Architect, Guy D. Channon.

1960 York: St. Peter's School. Big Hall extension. 1838 Architect, John Harper. Architects for extension, Needham, Thorp & White.

1961 Bedale: North Riding, Yorkshire. Alterations and repairs to Hornby Castle for Major W. E. Clutterbuck. Architects, Ward, Ruddick & Ward. Some medieval work, but mostly 18th century.

1961 Wakefield: Restoration of Heath Hall for M. M. Oddie. Architect, Francis F. Johnson. Built 1854 and designed by John Carr.

1961 Kendal: Westmorland. Abbott Hall. Grouting walls. 1759 Architect, John Carr; Engineer, John Dossor.

1962 Grantham: Lincs. Restoration of Belton House for Belton Trust Estate. Architect, Francis F. Johnson. 'Perhaps the most satisfying among the later 17th century houses in England.' (Pevsner.)

1962 Knaresborough: West Riding, Yorkshire. Alterations and repairs to Goldsborough Hall for F. Hanson. Architects, Taylor, Bown & Miller.

1962 Beverley: East Riding, Yorkshire. Restoration of south choir aisle roof to St. Mary's Church. Architect, George G. Pace.

1963 Everingham: East Riding, Yorkshire. Alterations and repairs to the Hall for the Trustees of Everingham Estate. Architect, Francis F. Johnson. Built 1757-60. Architect, John Carr.

1963 Whitby: North Riding, Yorkshire. Repairs to roofs and stonework of Mulgrave Castle for the Marquess of Normanby. Architects, Ward, Ruddick & Ward. An 18th century castellated house with additions by Sir John Soane and William Atkinson, and grounds by Humphrey Repton.

1963 Leeds: West Riding, Yorkshire. Church of St. Thomas, Stanningley. 1841 Architect, H. Rogerson. Repairs to tower and spire. Architects for the repairs, Charlton & Crowther.

1963 Cowthorpe: West Riding, Yorkshire. Medieval church: death watch beetle eradication to roof. Architects, Ward, Ruddick & Ward.

1964 Settrington: East Riding, Yorkshire. Restoration after fire of Settrington Hall for Lord Buckton. Architect, Francis P. Johnson. House built c. 1790.

1964 Kirk Deighton: West Riding, Yorkshire. Medieval church; extensive restorations 1875 W. Perkins: repairs to roofs. Architects for recent repairs, Charlton & Crowther.

1965 Tockwith: West Riding, Yorkshire. Skewkirk Hall. Additions and alterations for E. W. Fattorini, Esq. Architects, Ward, Ruddick & Ward.

WILLIAM PETER ANELAY

1965-72	Everingham: East Riding, Yorkshire. St. Everilda's Roman Catholic Church. Repairs to roof, etc. 1836 Architects, Agostino Georgioli & John Harper. Architect for the repairs, Francis F. Johnson.
1966	Coxwold: North Riding, Yorkshire. Alterations and repairs to Newburgh Priory for Captain V. M. Wombwell. Architects, Ward, Ruddick & Ward. Continued until 1969. Mostly 18th century, with some earlier work.
1966-7	Westow: East Riding, Yorkshire. The Hall, fine 18th century house in ashlar. Restoration for Lord Grimthorpe. Architect, Francis F. Johnson.
1966	Busby Hall: North Riding, Yorkshire. 1764 ashlar. Restoration for G. H. Marwood, Esq. Architects, Ward, Ruddick & Ward.
1967	York: Theatre Royal. Slate facings to the extension, designed by Patrick Gwynne.
1967	Bramham Park: West Riding, Yorkshire. Repairs to Mansion House and stables. Architects, Ward, Ruddick & Ward. The work continued until 1972.
1967	Keighley: West Riding, Yorkshire. Repairs to Tithe Barn, c. 1640, at East Riddlesden Hall for the National Trust. Architects, Hunter, Stoker and Mellor.
1967	York: Repairs to Merchant Adventurers' Hall. Architect, Francis F. Johnson. Buildings date from 16th century onwards.
1967, 1971 and 1973	York: Holy Trinity Church, Goodramgate. Restoration of medieval and 18th century church. Architects, Ward, Ruddick & Ward.
1968	Bramham Park: West Riding. Yorkshire, Repairs to Ponds and Cascades. Architects, Ward, Ruddick & Ward.
1968	Gawthorpe Hall: Lancs. Alterations and repairs for the Gawthorpe Trust. Architect, Francis F. Johnson. Built 1600-5. Drastic later work by Sir Charles Barry.
1968	York: Restorations to the Chapel of Red House School, Moor Monkton. Architect, George G. Pace. A second phase of work took place in 1971. Very interesting building of early 17th century date with fine furnishings, glass and murals.
1968	York: Stonework repairs to Yorkshire Insurance Offices, St. Helen's Square. Built 1840, designed by architect George Townsend Andrews.
1969	Whitby: North Riding, Yorkshire. Repairs to roof of St. Mary's Church. Architect Francis F. Johnson. Very interesting Norman and medieval church with unusual 18th century fittings.
1970	Ripon Cathedral: West Riding, Yorkshire. West forecourt and wood furniture for the Chapel of the Holy Spirit. Architects, Ward, Ruddick & Ward.
1970	York: Restoration of stonework, York County Savings Bank, St. Helen's Square. Built 1829-30, designed by architects Pritchett and Watson, with later work by E. Ridsdale Tate.
1970	Adel Church: West Riding, Yorkshire. Repairs. Architects, Charlton and Crowther. Interesting Norman, medieval, 18th and 19th century work, including 18th century glass.
1970	Helmsley: North Riding, Yorkshire. Woodward House. Alterations and restoration for the Woodward Society. Architects, Tom Adams Design Associates.
1970	Bishopthorpe: West Riding, Yorkshire. Repairs to Bishopthorpe Palace and Lodge for the Church Commissioners for England under the direction of the Commissioners' Chief Architect, R. G. Wood.
1971	Doncaster: Restoration of stonework at Cusworth Hall for Doncaster Rural District Council. Architect-in-charge, R. F. Haigh. Built 1740, designed by Architect George Platt and altered by James Paine.
1971	Leeds: Restorations at Kirkstall Abbey ruins for Leeds Corporation.
1971	Beverley: East Riding, Yorkshire. Restoration of north choir aisle roof and stonework of east gable of St. Mary's Church. Architect, George G. Pace.
1971-2	York: Castle Museum. 1705, probably designed by William Wakefield. Stonework repairs for York Corporation. E. Firth, City Architect.
1971	Tickhill Church: West Riding, Yorkshire. Great medieval church; repairs to chancel. Architect, George G. Pace.

1972 Durham: Restoration, Bishop Cosin's Library, University of Durham, Palace Green. Architect, George G. Pace. The buildings are of 16th, 17th and early 19th century date.

1972 Durham: Restoration of exterior of Tunstall Chapel, Durham Castle. Architect, George G. Pace. The part under repair includes 11th, 16th, 17th and early 19th century work.

1972 York: Assembly Rooms, 1730, Lord Burlington. 1828 portico added. Architects, Pritchett & Watson. Repairs to roofs and structure for York Corporation, R. H. Fogg, City Architect.

1972 York: Sir Thomas Herbert House. Repairs and alterations for Ings Property Company. Architects, Stansfield, Burton & Agar.

1972 Otley: West Riding, Yorkshire. Restoration of roofs, etc., at Weston Hall, for Colonel H. V. Dawson, under the supervision of Smiths Gore.

1973 North Dalton: East Riding, Yorkshire. Alterations, additions and restoration to Southwold House for the Marquess of Normanby. Architect, Francis F. Johnson.

1973 York: The Mansion House. Restoration to leadwork and roofs for York Corporation. R. H. Fogg, City Architect.

1973 Easingwold: North Riding, Yorkshire. Restoration work to the Church of St. John the Baptist. Architects, Stansfield, Burton & Agar.

1974 Studley Royal with Fountains Abbey Estate: West Riding, Yorkshire. Restoration of the Octagon Tower, Temple of Fame and gate piers to Fountains Hall for the West Riding County Council. Architect-in-charge, W. T. C. Walker, Deputy County Architect.

1974 York: 1 Museum Street. Stonework repairs to the North West elevations for York District Council. R. H. Fogg, City Architect.

1974 Bolton Percy: West Riding, Yorkshire. Restoration of the half-timbered, late 15th century Gatehouse. Architects, Brierley, Leckenby, Keighley and Groom.

1974 Northallerton: North Riding, Yorkshire. Restoration of roofs, gutters and turrets at Kiplin Hall for the trustees. Built c. 1625 by Lord Baltimore, founder of Maryland. Architects for the restoration, Ferrey & Mennim in association with Claude Phillimore.